LENS DESIGN FUNDAMENTALS

LENS DESIGN

FUNDAMENTALS

RUDOLF KINGSLAKE

Institute of Optics
University of Rochester
Rochester, New York

ACADEMIC PRESS, New York, San Francisco, London 1978

A Subsidiary of Harcourt Brace Jovanovich, Publishers

ACADEMIC PRESS, INC.
111 Fifth Avenue, New York, New York 10003

United Kingdom Edition published by
ACADEMIC PRESS, INC. (LONDON) LTD.
24/28 Oval Road, London NW1 7DX

Library of Congress Cataloging in Publication Data

Kingslake, Rudolf
    Lens design fundamentals.

    Includes bibliographical references.
    1. Lenses--Design and construction. I. Title.
QC385.2.D47K56   681'.42   77-80788
ISBN 0-12-408650-1

PRINTED IN THE UNITED STATES OF AMERICA

Contents

CHAPTER 4

Chromatic Aberration

CHAPTER 5

Spherical Aberration

CHAPTER 6

Design of a Spherically Corrected Achromat

CHAPTER 7

Oblique Pencils

CHAPTER 12

Symmetrical Double Anastigmats with Fixed Stop

CHAPTER 13

Unsymmetrical Photographic Objectives

CHAPTER 14

Mirror and Catadioptric Systems

CHAPTER 15

Eyepiece Design

CHAPTER 16

Automatic Lens Improvement Programs

Preface

This book can be regarded as an extension and modernization of Conrady's 50-year-old treatise, "Applied Optics and Optical Design," Part I of which was published in 1929.* This was the first practical text to be written in English for serious students of lens design, and it received a worldwide welcome.

It is obvious, of course, that in these days of rapid progress any scientific book written before 1929 is likely to be out of date in 1977. In the early years of this century all lens calculations were performed slowly and laboriously by means of logarithms, the tracing of one ray through one surface taking at least 5 minutes. Conrady, therefore, spent much time and thought on the development of ways by which a maximum of information could be extracted from the tracing of a very few rays. Today, when this can be performed in a matter of seconds or less on a small computer—or even on a programmable pocket calculator—the need for his somewhat complicated formulas has passed; but they remain valid and can be used profitably by any designer who takes the trouble to become familiar with them. In the same way, the third-order or Seidel aberrations have lost much of their importance in lens design. Even so, in some instances such as the predesign of a triplet photographic objective, third-order calculations still save an enormous amount of time.

Since Conrady's day, a great deal of new information has appeared, and new procedures have been developed, so that a successor to Conrady's book is seriously overdue. Many young optical engineers today are designing lenses with the aid of an optimization program on a large computer, but they have little appreciation of the how and why of lens behavior, particularly as these computer programs tend to ignore many of the classical lens types that have been found satisfactory for almost a century. Anyone who has had the experience of designing lenses by hand is able to make much better use of an optimization program than someone who has just entered the field, even though that newcomer may have an excellent academic background and be an expert in computer operation. For this reason an up-to-date text dealing with the classical processes of lens design will always be of value. The best that a computer can do is to optimize the system given to it, so the more understanding and competent the designer, the better the starting system he will be able to give the computer. A perceptive pre-

---

* A. E. Conrady, "Applied Optics and Optical Design," Part I, Oxford Univ. Press, London, 1929; also Dover, New York, 1957. Part II, Dover, 1960.

liminary study of a system will often indicate how many solutions exist in theory and which one is likely to yield the best final form.

A large part of this book is devoted to a study of possible design procedures for various types of lens or mirror systems, with fully worked examples of each. The reader is urged to follow the logic of these examples and be sure that he understands what is happening, noticing particularly how each available degree of freedom is used to control one aberration. Not every type of lens has been considered, of course, but the design techniques illustrated here can readily be applied to the design of other more complex systems. It is assumed that the reader has access to a small computer to help with the ray tracing, otherwise he may find the computations so time-consuming that he is liable to lose track of what he is trying to accomplish.

Conrady's notation and sign conventions have been retained, except that the signs of the aberrations have been reversed in accordance with current practice. Frequent references to Conrady's book have been given in footnotes as "Conrady, p. ..."; and as the derivations of many important formulas have been given by Conrady and others it has been considered unnecessary to repeat them here. In the last chapter a few notes have been added (with the help of Donald Feder) on the structure of an optimization program. This information is for those who may be curious to know what must go into such a program and how the data are handled.

This book is the fruit of years of study of Conrady's unique teaching at the Imperial College in London, of 30 years experience as Director of Optical Design at the Eastman Kodak Company, and of almost 45 years of teaching lens design in the Institute of Optics at the University of Rochester—all of it a most rewarding and never-ending education for me, and hopefully also for my students.

RUDOLF KINGSLAKE

LENS DESIGN FUNDAMENTALS

CHAPTER 1

The Work of the Lens Designer

Before a lens can be constructed it must be designed, that is to say, the radii of curvature of the surfaces, the thicknesses, the air spaces, the diameters of the various components, and the types of glass to be used must all be determined and specified. The reason for the complexity in lenses is that in the ideal case all the rays in all wavelengths originating at a given object point should be made to pass accurately through the image of that object point, and the image of a plane object should be a plane, without any appearance of distortion (curvature) in the images of straight lines.

Scientists always try to break down a complex situation into its constituent parts, and lenses are no exception. For several hundred years various so-called aberrations have been recognized in the imperfect image formed by a lens, each of which can be varied by changing the lens structure. Typical aberrations are spherical, chromatic, and astigmatic, but in any given lens all the aberrations appear mixed together, and correcting (or eliminating) one aberration will improve the resulting image only to the extent of the amount of that particular aberration in the overall mixture. Some aberrations can be easily varied by merely changing the shape of one or more of the lens elements, while others require a drastic alteration of the entire system.

The lens parameters available to the designer for change are known as "degrees of freedom." They include the radii of curvature of the surfaces, the thicknesses and airspaces, the refractive indices and dispersive powers of the glasses used for the separate lens elements, and the position of the "stop" or aperture-limiting diaphragm or lens mount. However, it is also necessary to maintain the required focal length of the lens at all times, for otherwise the relative aperture and image height would vary and the designer might end up with a good lens but not the one he set out to design. Hence each structural change that we make must be accompanied by some other change to hold the focal length constant. Also, if the lens is to be used at a fixed magnification, that magnification must be maintained throughout the design.

The word "lens" is ambiguous, since it may refer to a single element or to a complete objective such as that supplied with a camera. The term "system" is often used for an assembly of units such as lenses, mirrors, prisms, polarizers, and detectors. The name "element" always refers to a

1

single piece of glass having polished surfaces, and a complete lens thus contains one or more elements. Sometimes a group of elements, cemented or closely airspaced, is referred to as a "component" of a lens. However, these usages are not standardized and the reader must judge what is meant when these terms appear in a book or article.

I. RELATIONS BETWEEN DESIGNER AND FACTORY

The lens designer must establish good relations with the factory because, after all, the lenses that he designs must eventually be made. He should be familiar with the various manufacturing processes and work closely with the optical engineers. He must always bear in mind that lens elements cost money, and he should therefore use as few of them as possible if cost is a serious factor. Sometimes, of course, image quality is the most important consideration, in which case no limit is placed on the complexity or size of a lens. Far more often the designer is urged to economize by using fewer elements, flatter lens surfaces so that more lenses can be polished on a single block, lower-priced types of glass, and thicker lens elements since they are easier to hold by the rim in the various manufacturing operations.

A. SPHERICAL VERSUS ASPHERIC SURFACES

In almost all cases the designer is restricted to the use of spherical refracting or reflecting surfaces, regarding the plane as a sphere of infinite radius. The standard lens manufacturing processes[1] generate a spherical surface with great accuracy, but attempts to broaden the designer's freedom by permitting the use of nonspherical or "aspheric" surfaces lead to extremely difficult manufacturing problems; consequently such surfaces are used only when no other solution can be found. The aspheric plate in the Schmidt camera is a classic example. However, molded aspheric surfaces are very practical and can be used wherever the production rate is sufficiently high to justify the cost of the mold; this applies particularly to plastic lenses made by injection molding. Fairly accurate parabolic surfaces can be generated on glass by special machines.

In addition to the problem of generating and polishing a precise aspheric surface, there is the further matter of centering. Centered lenses with spherical surfaces have an optical axis that contains the centers of curvature of all the surfaces, but an aspheric surface has its own independent axis, which must be made to coincide with the axis containing all the other centers of

---

[1] F. Twyman, "Prism and Lens Making." Hilger and Watts, London, 1952. D. F. Horne, "Optical Production Technology." Crane Russak, New York, 1972.

curvature in the system. Most astronomical instruments and a few photographic lenses and eyepieces have been made with aspheric surfaces, but the designer is advised to avoid such surfaces if at all possible.

B. ESTABLISHMENT OF THICKNESSES

Negative lens elements should have a center thickness between 6 and 10% of the lens diameter, but the establishment of the thickness of a positive element requires much more consideration. The glass blank from which the lens is made must have an edge thickness of at least 1 mm to enable it to be held during the grinding and polishing operations (Fig. 1). At least 1 mm

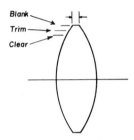

FIG. 1. Assigning thickness to a positive element.

will be removed in edging the lens to its trim diameter, and we must allow at least another 1 mm in radius for support in the mount. With these allowances in mind, and knowing the surface curvatures, the minimum acceptable center thickness of a positive lens can be determined. These specific limitations refer to a lens of average size, say $\frac{1}{2}$ to 3 in. in diameter; they may be somewhat reduced for small lenses, and they must be increased for large ones. A knife-edge lens is very hard to make and handle and it should be avoided wherever possible. A discussion of these matters with the glass-shop foreman can be very profitable.

As a general rule, weak lens surfaces are cheaper to make than strong surfaces because more lenses can be polished together on a block. However, if only a single lens is to be made, multiple blocks will not be used, and then a strong surface is no more expensive than a weak one.

A small point but one worth noting is that a lens that is nearly equiconvex is liable to be accidentally cemented or mounted back-to-front in assembly. If possible such a lens should be made exactly equiconvex by a trifling bending, any aberrations so introduced being taken up elsewhere in the system. Another point to notice is that a very small edge separation between two lenses is hard to achieve, and it is better either to let the lenses actually touch at a diameter slightly greater than the clear aperture, or to call for an edge separation of 1 mm or more, which can be achieved by a spacer ring or a rigid part of the mounting. Some typical forms of lens mount are shown in Fig. 2. Remember that the clearance for a shutter or an iris

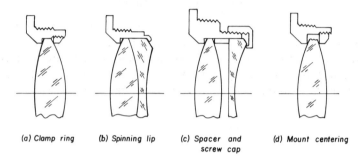

(a) Clamp ring   (b) Spinning lip   (c) Spacer and screw cap   (d) Mount centering

FIG. 2. Some typical lens mounts.

diaphragm must be counted from the bevel of a concave surface to the vertex of a convex surface.

C. ANTIREFLECTION COATINGS

Today practically all glass–air lens surfaces are given an antireflection coating to improve the light transmission and to eliminate ghost images. Since many lenses can be coated together in a large bell jar, the process is surprisingly inexpensive. However, for the most complete elimination of surface reflection over a wide wavelength range, a multilayer coating is required, and the cost then immediately rises.

D. CEMENTING

Small lens elements are often cemented together, using either Canada balsam or some suitable organic polymer. In lenses of diameter over about 3 in., however, the differential expansion of crown and flint glasses is liable to cause warpage or even fracture if a hard cement is used. Soft yielding cements or a liquid oil can be introduced between adjacent lens surfaces, but in large sizes it is more usual to separate the surfaces by small pieces of tinfoil or an actual spacer ring. The cement layer is always ignored in raytracing, the ray being refracted directly from one glass to the next.

The reasons for cementing lenses together are (a) to eliminate two surface reflection losses, (b) to prevent total reflection at the air film, and (c) to aid in mounting by combining two strong elements into a single, much weaker cemented doublet. The relative centering of the two strong elements is accomplished during the cementing operation rather than in the lens mount.

Cementing more than two lens elements together can be done, but it is very difficult to secure perfect centering of the entire cemented component. The designer is advised to consult with the manufacturing department before planning to use a triple or quadruple cemented component. Precise cementing of lenses is not a low-cost operation, and it is often cheaper to coat two surfaces that are airspaced in the mount rather than to cement these surfaces together.

E. Establishing Tolerances

It is essential for the lens designer to assign a tolerance to every dimension of a lens, for if he does not do so somebody else will, and that person's tolerances may be completely incorrect. If tolerances are set too loose a poor lens may result, and if too tight the cost of manufacture will be unjustifiably increased. This remark applies to radii, thicknesses, airspaces, surface quality, glass index and dispersion, lens diameters, and perfection of centering. These tolerances are generally found by applying a small error to each parameter, and tracing sufficient rays through the altered lens to determine the effects of the error. A knowledge of the tolerances on glass index and dispersion may make the difference between being able to use a stock of glass on hand, or the necessity of ordering glass with an unusually tight tolerance, which may seriously delay production and raise the cost of the lens. When making a single high-quality lens, it is customary to design with catalog indices, then order the glass, and then redesign the lens to make use of the actual glass received from the manufacturer. On the other hand, when designing a high-production lens, it is necessary to adapt the design to the normal factory variation of about ± 0.001 in refractive index and ± 0.2 in V value.

Matching thicknesses in assembly is a possible though expensive way to increase the manufacturing tolerances on individual elements. For instance, in a double-Gauss lens of the type shown in Fig. 3, the designer may deter-

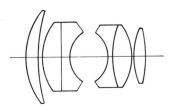

Fig. 3. A typical double-Gauss lens.

mine permissible thickness tolerances for the two cemented doublets in the following form:

each single element:	± 0.2 mm
each cemented doublet:	± 0.1 mm
the sum of both doublets:	± 0.02 mm

Clearly such a matching scheme requires that a large number of lenses be available for assembly, with a range of thicknesses. If every lens is made on the thick side no assemblies will be possible.

Very often the most important tolerances to specify are those for surface tilt and lens element decentration. A knowledge of these can have a great effect on the design of the mounting and on the manufacturability of the system. A decentered lens generally shows coma on the axis, whereas a tilted element often leads to a tilted field. Some surfaces are affected very little by a small tilt, whereas others may be extremely sensitive in this regard. A table of tilt coefficients should be in the hands of the optical engineers before they begin work on the mount design. The subject of optical tolerancing is almost a study in itself, and the setting of realistic tolerances is far from being an obvious or simple matter.

F. Design Tradeoffs

The lens designer is often confronted with a variety of ways to achieve a given result, and the success of a project may be greatly influenced by his choice. Some of these alternatives are as follows: Should a mirror or lens system be used? Can a strong surface be replaced by two weaker surfaces? Can a lens of high-index glass be replaced by two lenses of more common glass? Can an aspheric surface be replaced by two spherical surfaces? Can a long-focus lens working at a narrow angular field be replaced by a short-focus lens covering a wider field? Can a zoom lens be replaced by a series of normal lenses, giving a stepwise variation of magnification? If two lens systems are to be used in succession, how should the overall magnification be divided between them? Is it possible to obtain sharper definition if some unimportant aberration can be neglected?

II. THE DESIGN PROCEDURE

A closed mathematical solution for the constructional data of a lens in terms of its desired performance would be much too complex to be a real possibility. The best we can do is to use our knowledge of optics to set up a likely first approach to the desired lens, evaluate it, make judicious changes, reevaluate it, and so on. The process may be illustrated by a simple flow chart (Fig. 4). These four steps will be considered in turn.

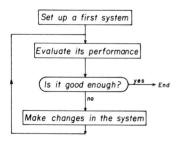

FIG. 4. Lens design flow chart.

A. SOURCES OF A LIKELY STARTING SYSTEM

In some cases, such as a simple telescope doublet, a lens design can be generated from first principles by a series of logical operations followed in a prescribed order. This is, however, exceptional. Far more often we obtain a likely starting system by one of the following means:

(1) A mental guess. This may work well for an experienced designer but it is hopeless for a beginner.

(2) A previously designed lens in the company files. This is the most usual procedure.

(3) Purchase of a competing lens and analysis of its structure. This is laborious and time-consuming, but it has often been done, especially in small firms with very little backlog of previous designs to choose from.

(4) A search through the patent files.

There are literally thousands of lens patents on file, but often the examples given are incomplete or not very well corrected; such a starting point may require a great deal of work before it is usable, not to mention the necessity of avoiding the claims in the patent itself! A recent book by Cox[2] includes an analysis of 300 lens patent examples, which the designer may find quite useful.

B. LENS EVALUATION

This is generally performed by tracing a sufficient number of rays

---

[2] A. Cox, "A System of Optical Design," p. 558. Focal Press, London and New York, 1964.

through the lens by accurate trigonometrical methods (see Chapter 2). At first only two or three rays are required, but as the design proceeds more rays must be added to provide an adequate evaluation of the system. There are several types of graph that can be plotted to represent the various aberrations, and a glance at these will often suggest to the designer what is wrong with the system.

C. LENS APPRAISAL

It is often very difficult to decide whether or not a given lens system is sufficiently well corrected for a particular application.[3] The usual method is to trace a large number of rays from a point source in a uniformly distributed array over the vignetted entrance pupil of the lens, and then plot a "spot diagram" of the points at which these rays pierce the image plane. It may be necessary to trace several hundred rays before a realistic appearance of the point image is obtained (see Chapter 7, Section IV). Chromatic errors can be included in the spot diagram by tracing sets of rays in several wavelengths, the spacing of the rays as they enter the lens being adjusted in accordance with the weight to be assigned to that wavelength in the final image.

To interpret the significance of a spot diagram, some designers calculate the diameters of circles containing 10, 20, 30, ..., 100% of the rays, and thus plot a graph of "encircled energy" at each obliquity. An alternative procedure is to regard the spot diagram as a point spread function, and by means of a Fourier transform convert it into a curve of MTF (modulation transfer function) plotted against spatial frequency. Such a graph contains very much information both as to the resolving power of the lens and the contrast in the image of coarse objects. Moreover, in calculating the MTF values, diffraction effects can be taken into account, the result being the most comprehensive representation of lens performance that can be obtained. If the lens is then constructed with dimensions agreeing exactly with the design data, it is possible to measure the MTF experimentally and verify that the lens performance has come up to the theoretical expectations.

D. SYSTEM CHANGES

When working by hand or with a small computer, the designer will have to decide what changes he should make to remove the residual aberrations in his lens. This is often a very difficult problem, and in the following chapters many hints are given as to suitable modifications that should be tried. Often a designer will make small trial changes in some of the lens parameters

---

[3] See, for instance, J. M. Palmer, "Lens Aberration Data." Elsevier, New York, 1971.

and determine the rate of change, or "coefficient," of each aberration with respect to each change. The solution of a few simultaneous equations will then indicate some reasonable changes that might be tried, although the extreme nonlinearity of all optical systems makes this procedure not as simple as one would like. Today there are many programs for use on a high-speed computer in which a large number of aberrations are changed simultaneously by varying several lens parameters, using a least-squares technique. In spite of the enormous amount of computation required in this process, it can be performed remarkably cheaply on today's large computers (see Chapter 16).

III. OPTICAL MATERIALS

The most common lens material is, of course, optical glass, but crystals and plastics are frequently used, while mirrors can be made of anything that is capable of being polished, even metals. Liquid-filled lenses have often been proposed, but for many obvious reasons they are practically never used. Optical materials in general have been discussed by Kreidl and Rood.[4]

A. OPTICAL GLASS

There are several well-known manufacturers of optical glass, and their catalogs give an enormous amount of information about the glasses that are available; in particular, the Schott catalog is virtually a textbook of optical glasses and their properties.

Optical glasses are classified roughly as crowns, flints, barium crowns, etc., but the boundaries of the various classes are not tightly standardized (see chart, Fig. 39, p. 78). Optically, glasses differ from one another in respect to refractive index, dispersive power, and partial dispersion ratio, while physically they differ in color, density, thermal properties, chemical stability, bubble content, striae, and ease of polishing.

Glasses vary enormously in cost, over a range of at least 300 to 1 from the densest lanthanum crowns to the most common ordinary plate glass, which is good enough for many simple applications. One of the lens designer's most difficult problems is how to make a wise choice of glass types, and in doing so he must weigh several factors. A high refractive index leads to weaker surfaces and therefore smaller aberration residuals, but high-index glasses are generally expensive, and they are also dense so that a pound of glass makes fewer lenses. If lens quality is paramount, then of course any

---

[4] N. J. Kreidl and J. L. Rood, Optical materials, in "Applied Optics and Optical Engineering" (R. Kingslake, ed.), Vol. I, pp. 153–200. Academic Press, New York, 1965.

glass can be used, but if cost is important the lower-cost glasses must be chosen. The cost of material in a small lens is likely to be insignificant, but in a large lens it may be a very serious matter, particularly as only a few types are made in large pieces (the so-called "massive optics"), and the price per pound is likely to vary as the cube of the weight of the piece. It is startling to note that in a lens of 12 in. diameter made of glass having a density of 3.5, each millimeter in thickness adds nearly 0.75 lb to the weight.

The color of glass is largely a matter of impurities, and some manufacturers offer glass with less yellow color at a higher price. This is particularly important if good transmission in the near ultraviolet is required. A trace of yellow color is often insignificant in a very small or a very thin lens and, of course, in aerial camera lenses yellow glass is quite acceptable because the lens will be used with a yellow filter anyway.

It will be found that the cost of glass varies greatly with the form of the pieces, whether in random slabs or thin rolled sheets, whether it is annealed, and whether it has been selected on the basis of low stria content. Some lens makers habitually mold their own blanks, and then it is essential to give them a long slow anneal to restore the refractive index to its stable maximum value; this is the value stated by the manufacturer on the melt sheet supplied with the glass.

B. INFRARED MATERIALS

Infrared-transmitting materials are a study in themselves, and many articles have appeared in the journals listing these substances and their properties. The Eastman Kodak Company[5] manufactures a line of hot-pressed polycrystalline materials known as Irtran®, which are summarized in the following tabulation. They are not generally usable in the visible, however, because of light scatter at the crystal boundaries.

Irtran no.	Material	Useful wavelength range (μm)	Approximate refractive index
1	Magnesium fluoride	0.5–9.0	1.36
2	Zinc sulfide	0.4–14.5	2.25
3	Calcium fluoride	< 0.4–11.5	1.42
4	Zinc selenide	0.5–22.0	2.44
5	Magnesium oxide	< 0.4–9.5	1.69
6	Cadmium telluride[a]	0.9–31.0	2.70

[a] J. E. Harvey and W. L. Wolfe, Refractive Index of Irtran 6 as a function of wavelength and temperature, J. Opt. Soc. Am., **65**, 1267 (1975).

[5] Publication U-72. Eastman Kodak Company, Rochester, New York.

Dispersion data on other useful infrared materials are given in the literature.

C. ULTRAVIOLET MATERIALS

For the ultraviolet region of the spectrum we have only fused quartz, calcium fluoride, and lithium fluoride, with a few of the lighter glasses when in thin sections. The situation is very unsatisfactory and mirrors are recommended in place of lenses, but even here there is a difficulty since most mirror coatings tend to reflect poorly in the ultraviolet.

D. OPTICAL PLASTICS

In spite of the paucity of available types suitable for lens manufacture, plastics have found extensive application in this field since World War I and particularly since the early 1950s.[6] Since that time at least 100 million plastic lenses have been fitted to inexpensive cameras, and they are now used regularly in eyeglasses and many other applications. Plastic triplets of $f/8$ aperture were first introduced by the Eastman Kodak Company in 1959, the " crown " material being methyl methacrylate and the " flint " a copolymer of styrene and acrylonitrile. The refractive indices of available optical plastics are very low, so that they fall into the region below the old crown–flint line, along with liquids and a few special titanium flints. The most common optical plastics are

	n_D	V
Methyl methacrylate:	1.49166	57.37
acrylonitrile–styrene copolymer:	1.56735	34.87
polystyrene:	1.59027	30.92

These dispersion data are not at all definite since they depend on such factors as the degree of polymerization and the temperature.

The advantages of plastic lenses are:

1. Ease and economy of manufacture in large quantities.
2. Low cost of the raw material.
3. The ability to mold the mount around the lens in one operation.
4. Lens thicknesses and airspaces are easier to maintain.
5. Aspheric surfaces can be molded as easily as spheres.
6. A dye can be incorporated in the raw material if desired.

---

[6] H. C. Raine, Plastic glasses, in "Proc. London Conf. Optical Instruments 1950" (W. D. Wright, ed.), p. 243. Chapman and Hall, London, 1951.

The disadvantages are:

1. The small variety and low refractive index of available plastics.
2. The softness of the completed lenses.
3. The high thermal expansion (eight times that of glass).
4. The high temperature coefficient of refractive index (120 times that of glass).
5. Plane surfaces do not mold well.
6. The difficulty of making a small number of lenses because of mold cost.
7. Plastics easily acquire high static charges, which pick up dust.
8. Plastic lenses cannot be cemented and can be coated only with difficulty.

In spite of these problems, plastic lenses have proved to be remarkably satisfactory in many applications, including low-cost cameras.

IV. INTERPOLATION OF REFRACTIVE INDICES

If we ever need to know the refractive index of an optical material for a wavelength other than those given in the catalog or used in measurement, some form of interpolation must be used, generally involving an equation connecting n with λ. A simple relation, which is remarkably accurate throughout the visible spectrum, is Cauchy's formula:[7]

$$n = A + B/\lambda^2 + C/\lambda^4$$

Indeed, the third term of this formula is often so small that when we plot n against $1/\lambda^2$ we obtain a perfectly straight line from the red end of the visible almost down to the blue-violet. For many glasses the curve is so straight that a very large graph may be plotted, and intermediate values picked off to about one in the fourth decimal place.

To use this formula, and the similar one due to Conrady,[8] namely,

$$n = A + B/\lambda + C/\lambda^{7/2} \tag{1}$$

it is necessary to set up three simultaneous equations for three known refractive indices and solve for the coefficients A, B, and C. In this way indices may be interpolated in the visible region to about one in the fifth decimal place.

---

[7] A. L. Cauchy, "Mémoire sur la Dispersion de la Lumière." J. G. Calve, Prague, 1936.

[8] A. E. Conrady, "Applied Optics and Optical Design," Part II, p. 659. Dover, New York, 1960.

Extrapolation is, however, not possible since the formulas break down beyond the red end of the spectrum.

Toward the end of the last century, several workers including Sellmeier, Helmholtz, Ketteler, and Drude, tried to develop a precise relationship between refractive index and wavelength based on resonance concepts.[9] The one most generally employed is

$$n^2 = A + \frac{B}{\lambda^2 - C^2} + \frac{D}{\lambda^2 - E^2} + \frac{F}{\lambda^2 - G^2} + \cdots \qquad (2)$$

In this formula the refractive index becomes infinite when λ is equal to C, E, G, etc., so that these values of λ represent asymptotes marking the centers of absorption bands. Between asymptotes the refractive index follows the curve indicated schematically in Fig. 5.

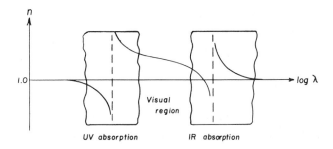

FIG. 5. Schematic relation between the refractive index of a glass and the log of the wavelength.

For most glasses and other transparent uncolored media, two asymptotes are sufficient for interpolation purposes, one representing an ultraviolet absorption and the other an infrared absorption. The visible spectrum is then covered by values of λ lying between the two absorption bands.

Expanding Eq. (2) by the binomial theorem, we obtain an approximate form of this equation, namely,

$$n^2 = a\lambda^2 + b + c/\lambda^2 + d/\lambda^4 + \cdots$$

in which the coefficient a controls the infrared indices (large λ) while coefficients c, d, etc., control the ultraviolet indices (small λ). If the longer infrared is of importance in some particular application, then it is advisable to add one or more terms of the type $e\lambda^4 + f\lambda^6$, etc.

---

[9] See, for instance, P. Drude, "The Theory of Optics," p. 391. Longmans Green, New York and London, 1922.

Herzberger[10] has proposed a somewhat[11] different formula, namely,

$$n = A + B\lambda^2 + \frac{C}{\lambda^2 - \lambda_0^2} + \frac{D}{(\lambda^2 - \lambda_0^2)^2}$$

in which A, B, C, D are coefficients for any given glass, and λ_0 has a fixed value for all glasses. He found that a suitable value is given by $\lambda_0^2 = 0.035$, or $\lambda_0 = 0.187$. This takes care of the ultraviolet absorption, and the near infrared is covered by the $B\lambda^2$ term. If the infrared is more important, another infrared term should be added.

In the current Schott glass catalog a six-term expression has been used for smoothing the stated index data. It is

$$n^2 = A_0 + A_1\lambda^2 + A_2/\lambda^2 + A_3/\lambda^4 + A_4/\lambda^6 + A_5/\lambda^8$$

which provides a very high degree of control in the blue and ultraviolet regions, but it is not valid much beyond 1 μm in the infrared.

The Bausch and Lomb Company[12] has adopted the following seven-term formula for their interpolation:

$$n^2 = a + b\lambda^2 + c\lambda^4 + \frac{d}{\lambda^2} + \frac{e\lambda^2}{(\lambda^2 - f) + g\lambda^2/(\lambda^2 - f)}$$

This is an awkward nonlinear type of relationship involving a considerable computing problem to determine the seven coefficients for any given type of glass.

A. INTERPOLATION OF DISPERSION VALUES

When using the $(D - d)$ method of achromatism (p. 93), it is necessary to know the Δn values of the various glasses for the particular spectral region that is being used. For achromatism in the visible, the Δn is usually taken to be $(n_F - n_C)$, but for any other spectral region a different value of Δn must be used. Indeed, a change in the relative values of Δn is really the only factor that determines the spectral region for the achromatism.

To calculate Δn we must differentiate the (n, λ) interpolation formula. This gives us the value of $dn/d\lambda$, which is the slope of the (n, λ) curve at any particular wavelength. The desired value of Δn is then found by multiplying

---

[10] M. Herzberger, Colour correction in optical systems and a new dispersion formula, *Opt. Acta (London)* **6**, 197 (1959).

[11] M. Herzberger, "Modern Geometrical Optics," p. 121. Wiley (Interscience), New York, 1958.

[12] N. J. Kreidl and J. L. Rood, ref. 4, p. 161.

$(dn/d\lambda)$ by a suitable value of $\Delta\lambda$. Actually, the particular choice of $\Delta\lambda$ is unimportant since we shall be working toward a zero value of $\sum (D - d) \Delta n$, but if we are expecting to compare a residual of $\sum (D - d) \Delta n$ with some established tolerance, it is necessary to adopt a value of $\Delta\lambda$ that will yield a Δn having approximately the same magnitude as the $(n_F - n_C)$ of the glass.

As an example, suppose we are using Conrady's interpolation formula, and we wish to achromatize a lens about some given spectral line. Then by differentiating Eq. (1) we get

$$\frac{dn}{d\lambda} = -\frac{b}{\lambda^2} - \frac{7}{2}\frac{c}{\lambda^{9/2}} \tag{3}$$

This formula contains the b and c coefficients of the particular glass being used, and also the wavelength λ at which we wish to achromatize, say, the mercury g line.

Suppose we are planning to use Schott's SK-6 and SF-9 types. Solving Eq. (1) for two known wavelengths, we find

Glass	b	c	$dn/d\lambda$ at the g line
SK-6	0.0124527	0.000520237	-0.142035
SF-9	0.0173841	0.001254220	-0.275885

For wavelength 0.4358 μm we find for these two glasses that $\Delta n = 0.010369$ and 0.020140, respectively, using the arbitrary value of $\Delta\lambda = -0.073$. These values should be compared with the ordinary $\Delta n = (n_F - n_C)$ values, which for these glasses are 0.01088 and 0.01945 respectively. It is seen that the flint dispersion has increased relative to the crown dispersion, which is characteristic of the blue end of the spectrum.

B. TEMPERATURE COEFFICIENT OF REFRACTIVE INDEX

If the ambient temperature in which the lens is to be used is liable to vary greatly, we must consider the resulting change in the refractive indices of the materials used. For glasses this usually presents no problem since the temperature coefficient of refractive index is very small, of the order of 0.000001 per 1°C.[13] However, for crystals it is likely to be much greater, and for plastics it is very large:

fluorite   0.00001 per 1°C

plastics   0.00014 per 1°C

---

[13] F. A. Molby, Index of refraction and coefficients of expansion of optical glasses at low temperatures, *J. Opt. Soc. Am.* **39**, 600 (1949).

Thus over a normal temperature range, say from 0 to 40°C, the refractive index of plastic lenses changes by 0.0056, quite enough to alter the focus significantly. In a reflex camera this would be overcome during the focusing operation before making the exposure, but in a fixed-focus or rangefinder camera, or one depending on the use of a focus scale, something must be done to avoid this temperature effect. One way that has been proposed is to place all or most of the lens power in a glass element, using the plastic elements only for aberration correction. Another suggestion is to mount the lens in a compensated mount of two materials having very different coefficients of expansion, so that as the temperature changes, one airspace of the lens is altered by just the right amount to restore the image position on the film. The thermal expansion of plastics is also large, but this is immaterial if the camera body is also made of plastic, since a temperature change then merely expands or contracts the entire apparatus, leaving the image always in the same plane.

V. LENS TYPES TO BE CONSIDERED

Lenses fall into several well-defined and well-recognized types, many of which will be considered in this book. They are

1. Lenses giving excellent definition only on axis
   a. Telescope doublets (low aperture)
   b. Microscope objectives (high aperture)
2. Lenses giving good definition over a wide field
   a. Photographic objectives
   b. Projection lenses
   c. Flat-field microscope objectives
3. Lenses covering a finite field with a remote stop
   a. Eyepieces, magnifiers, and loupes
   b. Viewfinders
   c. Condensers
   d. Afocal Galilean or anamorphic attachments
4. Catadioptric (mirror–lens) systems
5. Varifocal and zoom lenses

Each of these types, and indeed every form of lens, requires an individual and specific process for its design. Some lenses contain many refracting surfaces while some contain few. In some lenses there are so many available parameters that almost any glass can be used; in others the choice of glass is an important degree of freedom. Some lens systems favor a high relative aperture but cover only a small angular field, while other types are just the reverse.

Several classical lens types are considered in this book and the design of a specific example of each is shown in detail. The student is strongly advised to follow through these designs carefully, since they employ a number of well-recognized techniques that can often be usefully applied to other design situations.

Some of the procedures that have been utilized in the examples in this book are:

1. Lens bending
2. Shift of power from one element to another
3. Single and double graphs, to vary one or two lens parameters simultaneously
4. Symmetry, for the automatic removal of the transverse aberrations
5. Selection of stop position by the $(H' - L)$ plot
6. Achromatism by the $(D - d)$ method
7. Selection of glass dispersions at the end of a design
8. The matching principle for the design of a high-aperture aplanat
9. Use of a "buried surface" for achromatism
10. Reduction of Petzval sum, by the three classical methods
11. Use of a narrow airspace to reduce zonal spherical aberration
12. Introduction of vignetting to cut off bad rim rays
13. Solution of four aberrations by the use of four simultaneous equations.

Since this book is primarily directed toward the needs of the beginner, no reference has been made to the more complex modern photographic objectives. This omission includes particularly high aperture lenses of the double-Gauss and Sonnar types, and wide-angle lenses such as the Biogon and reversed telephoto. Zoom lenses and afocal and anamorphic attachments have been omitted for the same reason. Today these complex systems are invariably designed with the aid of an optimization program on a large computer.

CHAPTER 2

Meridional Ray Tracing

I. INTRODUCTION

It is reasonable to assume that anyone planning to study lens design is already familiar with the basic facts of geometrical and physical optics. However, there are a few points that should be stressed to avoid confusion or misunderstanding on the part of the reader.

A. Object and Image

All lens design procedures are based on the principles of geometrical optics, which assumes that light travels along rays that are straight in a homogeneous medium. Light rays are refracted or reflected at a lens or mirror, whence they proceed to form an image. Due to the inherent properties of refracting and reflecting surfaces and the dispersion of refracting media, the image of a point is seldom a perfect point but is generally afflicted with aberrations. Further, owing to the wave nature of light, the most perfect image of a point is always, in fact, a so-called Airy disk, a tiny patch of light of the order of a few wavelengths in diameter surrounded by decreasingly bright rings of light.

It should be remembered that both objects and images can be either "real" or "virtual." The object presented to the first surface of a system is, of course, always real. The second and following surfaces may receive converging or diverging light, indicating respectively a virtual or real object for that surface. It must never be forgotten that in either case the refractive index to be applied to the calculation is that of the space containing the entering rays at the surface under consideration. This is known as the object space for that surface.

Similarly, the space containing the rays emerging from a surface is called the image space, and real or virtual images are considered to lie in this space. Because of the existence of virtual objects and virtual images we must regard the object and image spaces as overlapping to infinity in both directions.

B. The Law of Refraction

The well-known Snell's law is generally written

$$n' \sin I' = n \sin I$$

where I, I' are, respectively, the angles between the incident and refracted rays and the normal at the point of incidence, while n, n' are the refractive indices of the media containing the incident and refracted rays, respectively. Refractive index is the ratio of the velocity of light in air to its velocity in the medium, and the refractive indices of all transparent media vary with wavelength, being greater for blue light than for red. The refractive index of vacuum relative to air is about 0.9997, which must occasionally be taken into account if a lens is to be used in vacuum.

The second part of the law of refraction is that the incident ray, the refracted ray, and the normal at the point of incidence all lie in one plane, called the plane of incidence. This part of the law becomes of importance in the tracing of skew rays (see p. 147).

For reflection we merely write $n' = -n$; this is because I' at a mirror surface is equal to I but with opposite sign. Thus, if a clockwise rotation takes us from the normal to the incident ray, it will require an equal counterclockwise rotation to go from the normal to the reflected ray.

C. The Meridian Plane

In this book we shall consider only centered systems, that is, lenses in which the centers of curvature of spherical surfaces, and the axes of symmetry of aspheric surfaces, all lie on a single optical axis. An object point lying on this axis is called an axial object, while one lying off-axis is called an extraaxial object point. The plane containing an extraaxial object point and the lens axis is known as the meridian plane; it constitutes a plane of symmetry for the whole system.

D. Types of Rays

If the object point lies on the lens axis we trace only axial rays. However, for an extraaxial object point there are two kinds of rays to be traced, namely, meridional rays, which lie in the meridian plane, shown in the familiar ray diagram of a system, and skew rays, which lie in front of or behind the meridian plane and do not intersect the axis anywhere. Each skew ray pierces the meridian plane at the object point and also at another point in the image space known as the diapoint of the ray. The paths of two typical skew rays are shown diagrammatically in Fig. 6.

Axial rays and meridional rays can be traced by relatively simple trigonometric formulas, or even graphically if very low precision is adequate. Skew rays, on the other hand, are much more difficult to trace, the procedure being discussed in Chapter 7, Section III.

An important limiting class of rays that have many applications are the so-called paraxial rays, which lie throughout their length so close to the

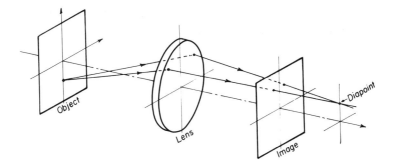

FIG. 6. A typical pair of skew rays.

optical axis that their aberrations are negligible. The ray tracing formulas for paraxial rays contain no trigonometrical functions and are therefore well suited to algebraic manipulation. A paraxial ray is really only a mathematical abstraction, for if the diaphragm of a real lens were stopped down to a very small aperture in an effort to isolate only paraxial rays, the depth of focus would become so great that no definite image could be located, although the theoretical image position can be calculated as a mathematical limit.

For an oblique ray in the meridian plane it is useful to consider two limiting rays very close to the traced ray, one slightly above or below it in the meridian plane, and the other a sagittal (skew) ray lying just in front of or behind the traced ray. These are used in the calculation of astigmatism (see Chapter 10, Section I).

E. NOTATION AND SIGN CONVENTIONS

This is a very vexed subject, as every lens designer has his own preferred system, which never seems to agree with that used by others. In spite of the efforts of several committees that have been appointed since World War II, no standard system has been established. In this book we adhere strictly to Conrady's notation except for the signs of the aberrations. In Conrady's day it was customary to regard all the properties of a single positive lens as positive, whereas today it is universal to regard undercorrected aberrations as negative and overcorrected aberrations as positive. This change in the prevailing attitude leads to a reversal of the sign of all Conrady's aberration expressions, requiring care on the part of any reader who is familiar with the earlier writings on practical optics.

So far as meridional rays are concerned, the origin of coordinates is placed at the vertex of a refracting or reflecting surface, with distances measured along the axis (the X axis) as positive to the right and negative to the left of this origin (Fig. 7). Transverse distances Y in the meridian plane

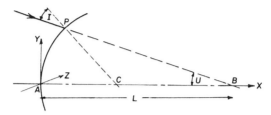

FIG. 7. A typical meridional ray incident on a spherical surface (all-positive diagram).

are considered positive if above the axis and negative below it. For skew rays, distances Z in the third dimension perpendicular to the meridian plane are generally considered positive when behind that plane, because then the Y and Z dimensions occupy their normal directions when viewed from the image space looking back into the lens. However, in a centered system all Z dimensions are symmetrical about the meridian plane, so that any phenomenon having a $+Z$ dimension is matched by a similar phenomenon having an identical $-Z$ dimension, as if the whole of the Z space were imaged by a plane mirror lying in the meridian plane itself.

For the angles, we shall regard the slope U of a meridional ray as positive if a clockwise rotation takes us from the axis to the ray, and the angle of incidence I as positive if a counterclockwise rotation takes us from the normal to the ray. These angle conventions are admittedly inconsistent, and there is a strong move at the present time to reverse the sign of U. Unfortunately this change leads to the introduction of as many minus signs as it removes, and worse still, it becomes impossible to draw an all-positive diagram for use when deriving computing formulas. In Conrady's system the paraxial ray height y is equal to (lu), but in the proposed new system this becomes $(-lu)$. The presence of these negative signs is a great inconvenience, and we shall therefore continue to use Conrady's angle conventions. Of course, the reader can always reverse the sign of U and u wherever they occur if he so wishes.

Finally, all data relating to the portion of a ray lying in the space to the left of a surface, usually the object space, are represented by unprimed symbols, while data referring to the portion of a ray lying in the space to the right of a surface are denoted by primed symbols. In a mirror system where the object and image spaces overlap, data of the entering ray are unprimed while those of the reflected ray are primed, even though both rays lie physically on the same side of the mirror. Mirror systems are considered fully in Chapter 14.

II. GRAPHICAL RAY TRACING

For many purposes, such as in the design of condenser lenses, a graphical ray trace is entirely adequate. The procedure is based on Snell's construction; it has been described by Dowell and van Albada.[1] It is illustrated in

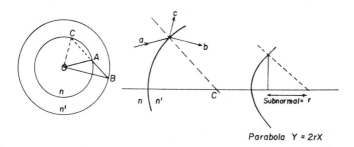

Fig. 8. Graphical ray tracing.

Fig. 8. Having made a large-scale drawing of the lens, we add a series of concentric circles at any convenient place on the paper about a point O, of radii proportional to the refractive indices of all the materials in the system. A convenient scale for these circles is 10 cm radius for air and 16 cm radius for a glass of index 1.6.

Having drawn the incident ray on the lens diagram, a line is drawn through O parallel to the incident ray to cut at A the index circle corresponding to the index of the medium containing the incident ray. A line is next drawn through A parallel to the normal at the point of incidence, to cut the circle corresponding to the index of the next medium at B; then OB will be the direction of the refracted ray in the medium B. This process is repeated for each refracting surface in the system. Mirrors can be handled by drawing the normal line right across the diagram to intersect the same index circle on the opposite side (point C). It is convenient to draw the index circles in ink, and to indicate rays by little pencil marks labeled with the same letters as the rays on the lens diagram. Some workers make a practice of erasing each mark after the next mark has been made, to avoid confusion. System changes can be made conveniently by laying a sheet of tracing paper over the diagram and marking the changes on the new paper; this permits the previous system to be seen as well as the changes.

A ray can be traced graphically through an aspheric surface if the direc-

---

[1] J. H. Dowell, Graphical methods applied to the design of optical systems, in *Proc. Opt. Convention 1926*, p. 965. See also L. E. W. van Albada, "Graphical Design of Optical Systems." Pitman, London, 1955.

tion of the normal is known. A parabolic surface is particularly simple, since the subnormal of a parabola is equal to the vertex radius (Fig. 8). Graphical ray tracing is rapid and easy, and at any time the ray can be traced accurately by trigonometry to confirm the graphical trace. It also enables the designer to keep track of the lens diameters and thicknesses as he moves along. A more complicated graphical ray trace ascribed to Thomas Young is given on p. 191. Paraxial rays can also be traced graphically (see p. 44).

III. TRIGONOMETRICAL RAY TRACING AT A SPHERICAL SURFACE

The path of a meridional ray through a single spherical refracting surface can be traced with high accuracy by various well-established procedures that will now be described. The ray emerging from one surface is then transferred to the next surface, where the whole process is repeated until the ray emerges into the final image space.

A. The (L, U) Method

For many years, especially when logarithms were used for calculation, it was customary to define a ray in relation to a surface by stating the slope angle U and the intersection length L. The procedure is described clearly in Conrady's book.[2] However, this method is seldom used today because L can become infinite, and in any case the method breaks down for a long radius or a plane surface. Neither of these possibilities is acceptable when writing a computer program for tracing rays, and consequently today other methods are invariably employed.

B. The (Q, U) Method

Here the ray is defined in relation to the surface by stating the ray slope angle U and the perpendicular distance Q of the ray from the surface vertex, the sign conventions being as described in Section I,E. To derive the ray tracing equations, we consider Fig. 9a, in which a line has been drawn parallel to the ray through the center of curvature of the surface. This line divides Q into two parts; hence $Q = r \sin I + r \sin U$. From this relationship we find

$$\sin I = Qc - \sin U \tag{4a}$$

────────────

[2] A. E. Conrady, p. 7.

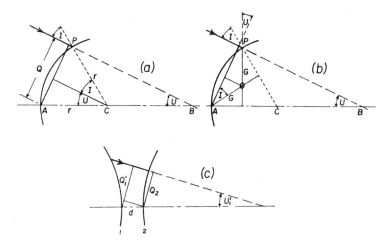

FIG. 9. Development of the ray tracing equations. (a) A short radius; (b) universal; (c) the transfer.

where c is the curvature of the surface, the reciprocal of the radius of curvature.

Now by the law of refraction we determine the angle of refraction as

$$\sin I' = (n/n') \sin I \qquad (4b)$$

and since angle PCA is equal to $U + I$ and $U' + I'$, evidently

$$U' = U + I - I' \qquad (4c)$$

Finally, we calculate the Q' of the refracted ray by placing primes on all the terms in Eq. (4a), giving

$$Q' = (\sin U' + \sin I')/c \qquad (5)$$

This equation is good when the radius of curvature is fairly short so that c is large, but at a weak surface of long radius, I' approaches $-U'$, so that Q' becomes the ratio of two small numbers; while if the surface is plane, Q' is actually $0/0$ and is indeterminate. Consequently other equations have been developed to replace Eq. (5). In Fig. 9b we draw a perpendicular from the point of incidence P to the lens axis and another perpendicular from the surface vertex A to the normal. These perpendiculars intersect at O, and by similar triangles we see that $OA = OP = G$, say. We now draw a line parallel to the ray through O to divide Q into two parts, the upper part being $(G \cos U)$ and the lower part $(G \cos I)$. Hence $Q = G(\cos U + \cos I)$, and

$$G = Q/(\cos U + \cos I)$$

The virtue of this relation is that cosines are always positive, and hence G is always about equal to one-half of Q, more or less. Under no circumstances can this expression become indeterminate. By adding primes everywhere, we reach the final relation:

$$Q' = G(\cos U' + \cos I') \qquad (6)$$

For a plane surface, a little consideration shows that

$$\sin U' = (n/n') \sin U, \qquad Q' = Q \cos U'/\cos U$$

1. Transfer

The transfer of the ray from one surface to the next is clear from Fig. 9c, in which

$$Q_2 = Q'_1 - d \sin U'_1$$

2. Right-to-Left Calculations

The regular equations can be used for tracing a ray in a right-to-left direction, but now we are given U' and Q' and we must determine U and Q. In the transfer we must now *add* ($d \sin U_2$) to Q_2 to get Q'_1. Remember that plain symbols refer to sections of rays lying to the left of the surface, and primed symbols to ray sections lying to the right of the surface.

3. Summary

Since we sometimes need to know the X, Y coordinates of the point of incidence of a ray at a surface, suitable equations are added at the end of the following summary.

The meridional ray tracing equations may be summarized as follows:

$$\sin I = Qc - \sin U$$
$$\sin I' = (n/n') \sin I \qquad (7)$$
$$U' = U + I - I'$$

Short radius only:

$$Q' = (\sin I' + \sin U')/c$$

Universal:

$$G = Q/(\cos U + \cos I)$$
$$Q' = G(\cos U' + \cos I')$$

Opening:

$$\text{initial } Q = L \sin U$$

Transfer:

$$Q_2 = Q'_1 - d \sin U'_1$$

Closing:

$$L' = (\text{final } Q')/(\text{final } \sin U')$$

Point of incidence:

$$Y = \sin(U + I)/c, \qquad Y = G[1 + \cos(U + I)]$$
$$X = [1 - \cos(U + I)]/c, \qquad X = G \sin(U + I) \tag{8}$$

Throughout this book it is anticipated that calculations will be performed on a small pocket electronic calculator in which sines and arc sines are given to eight or ten significant figures at the touch of a button. Only some of the computed quantities need be recorded, therefore, and angles will be stated to five decimals of a degree, or 1/28 sec of arc. Obviously this precision is much higher than that to which optical parts can be manufactured, but since we often calculate aberrations as the small difference between two very nearly equal large numbers, this extra precision is quite useful.

4. *Special Cases*

There are two special cases that should be recognized:

a. If $\sin I$ is greater than 1.0, this indicates that the radius is so short that the ray misses the surface altogether.

b. If $\sin I'$ is greater than 1.0, this indicates total internal reflection.

For manual work a suitable arrangement of the calculation is shown in Table I. This represents the tracing of a marginal ray entering parallel to the axis at a height of 2.0 into a cemented doublet of focal length 12, having the following structure (Fig. 10):

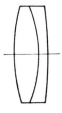

FIG. 10. A cemented doublet objective.

$$r_1 = 7.3895, \qquad c_1 = 0.135327,$$
$$r_2 = -5.1784, \qquad c_2 = -0.19311, \qquad \begin{array}{ll} d'_1 = 1.05, & n'_1 = 1.517 \\ d'_2 = 0.40, & n'_2 = 1.649 \end{array}$$
$$r_3 = -16.2225, \qquad c_3 = -0.06164$$

Note that in this table of computation the sine of the ray slope is written above the angle U itself, since it is required twice more, for the transfer and again for the calculation of $\sin I$ at the following surface. It is also worth recording the $\cos U$ below the angle since it is required for the G calculation. The X and Y values have been added at the end of each surface, although they are not always required. Data relating to the surfaces appear in the columns, while data referring to the spaces between surfaces appear on the vertical dividing lines between the surface data.

TABLE I

TRACING OF A MARGINAL RAY THROUGH A CEMENTED DOUBLET

$c = 1/r$		0.1353271		−0.1931098		−0.0616427	
d			1.05		0.40		
n			1.517		1.649		
Q		2.0		1.9178334		1.9186619	
Q'		2.0171179		1.9398944		1.8814033	
I		15.70320°		−27.70433°		−9.86648°	
I'		10.27740°		−25.32129°		−16.41309°	
$\sin U$	0		0.0945566		0.0530812		0.1665858
U	0		5.42580°		3.04276°		9.58937°
$\cos U$			0.9955195		0.9985902		
G		1.0190166		1.0196480		0.9671650	
Y		2.0		1.9631812		1.9274790	
X		0.2758011		−0.3865582		−0.1149137	

$$\text{Closing:} \qquad L = \frac{\text{final } Q'}{\text{final } \sin U'} = 11.293900$$

C. PROGRAM FOR A COMPUTER

When programming this procedure for a computer, it is of course possible to use available sine and arc sine subroutines, but it is generally much quicker to work through the square root, remembering

$$\sin(a + b) = \sin a \cos b + \cos a \sin b$$

and

$$\cos(a + b) = \cos a \cos b - \sin a \sin b$$

Given Q, $\sin U$, and $\cos U$, the equations to be programmed are

$$\sin I = Qc - \sin U$$

$$\cos I = (1 - \sin^2 I)^{1/2}$$

$$\left.\begin{array}{l} \sin(U + I) = \sin U \cos I + \cos U \sin I \\ \cos(U + I) = \cos U \cos I - \sin U \sin I \end{array}\right\} \quad \text{(A)}$$

$$\sin(-I') = -(n/n') \sin I$$

$$\cos I' = [1 - \sin^2(-I')]^{1/2}$$

$$\left.\begin{array}{l} \sin U' = \sin(U + I) \cos(-I') + \cos(U + I) \sin(-I') \\ \cos U' = \cos(U + I) \cos(-I') - \sin(U + I) \sin(-I') \end{array}\right\} \quad \text{(B)}$$

$$G = Q/(\cos U + \cos I)$$

$$Q' = G(\cos U' + \cos I')$$

Transfer:

$$Q_2 = Q_1' - d \sin U_1'$$

Note: the three equations in (A) and (B) are identical with different numbers substituted. It is therefore convenient to write a "cosine cross-product subroutine" to handle the three equations, and substitute the appropriate numbers each time it is used. Remember, of course, that the cosine of a negative angle is positive. When using this routine it is necessary to carry over both $\sin U'$ and $\cos U'$ to become $\sin U$ and $\cos U$ at the next surface.

IV. SOME USEFUL RELATIONS

A. The Spherometer Formula

The relation between the height Y and the sag X of a spherical surface of radius r is often required. It is evident from Fig. 11 that $r^2 = Y^2 + (r - X)^2$; hence

$$X = (X^2 + Y^2)/2r = r - (r^2 - Y^2)^{1/2}$$

This can be expanded by the binomial theorem to give

$$X = \frac{Y}{2}\left(\frac{Y}{r}\right) + \frac{Y}{8}\left(\frac{Y}{r}\right)^3 + \frac{Y}{16}\left(\frac{Y}{r}\right)^5 + \cdots \tag{9}$$

Because r can become infinite, it is generally better to express X in terms of the surface curvature c rather than the radius r. Writing $c = 1/r$ gives

$$X = cY^2/[1 + (1 - c^2Y^2)^{1/2}] \tag{10}$$

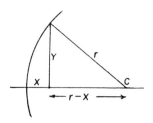

FIG. 11. The spherometer formula.

This expression never becomes indeterminate. For a plane surface, $c = 0$ and of course $X = 0$ also.

B. SOME USEFUL FORMULAS

There are a number of useful relations that can be readily derived between the quantities involved in ray tracing at a spherical surface. Some of them are

$$G = r \tan \tfrac{1}{2}(U + I) = PA^2/2Y$$

$$(\text{chord}) PA = 2r \sin \tfrac{1}{2}(U + I) = 2G \cos \tfrac{1}{2}(U + I)$$

$$Y = PA \cos \tfrac{1}{2}(U + I) = PA^2(\cos U + \cos I)/2Q$$

$$X = PA \sin \tfrac{1}{2}(U + I) = PA^2(\sin U + \sin I)/2Q$$

$$X = Y \tan \tfrac{1}{2}(U + I) = Y(\sin U + \sin I)/(\cos U + \cos I)$$

The following relations also involve the refraction of a ray at a surface:

$$n' \sin I' - n \sin I = Y \left[\frac{n' \cos U' - n \cos U}{r - X} \right]$$

$$= Y \left[\frac{n' \cos I' - n \cos I}{r} \right]$$

$$n'L \sin U' - nL \sin U = r(n' \sin U' - n \sin U) = n'Q' - nQ$$

$$n' \cos U' - n \cos U = \cos(U + I)(n' \cos I' - n \cos I)$$

$$\tan \tfrac{1}{2}(I + I') = \tan \tfrac{1}{2}(I - I')(n' + n)/(n' - n)$$

C. THE INTERSECTION HEIGHT OF TWO SPHERES

If we decide to make two lenses touch at the edge as an aid to mounting, we must choose an axial separation such that the two adjacent surfaces intersect at a diameter lying between the clear aperture and trim diameter of the lenses. Or again, if we wish to reduce the thickness of a large lens to its absolute minimum, we must be able to calculate the thickness so that the lens surfaces intersect at the desired diameter, plus a small addition to provide sufficient edge thickness.

Given r_1, r_2, and the axial thickness d, we first calculate

$$A = (2r_2 + d)/(2r_1 - d)$$

Then $X_2 = d/(A - 1)$ and $X_1 = AX_2 = (X_2 + d)$, and the intersection height Y is given by

$$Y = (2r_1 X_1 - X_1^2)^{1/2} = (2r_2 X_2 - X_2^2)^{1/2}$$

Example. If $r_1 = 50$, $r_2 = 250$, and $d = 3$, we find that $A = 503/97 = 5.18556$. Then $X_2 = 0.71675$ and $X_1 = 3.71675$, giving $Y = 18.917$.

D. THE VOLUME OF A LENS

To calculate the volume of a lens, and hence its weight, we divide the lens into three parts, the two outer spherical "caps" and a central cylinder. The volume of each of the caps is found by the standard formula

$$\text{volume} = \tfrac{1}{3}\pi X^2(3r - X)$$

or, by eliminating r, we have

$$\text{volume} = \tfrac{1}{2}\pi Y^2 X + \tfrac{1}{6}\pi X^3 \tag{11}$$

For many purposes only the first term of (11) need be used, showing that the "average" thickness of the cap is approximately $\tfrac{1}{2}X$. The lens thus has the volume of a cylinder of thickness $\tfrac{1}{2}X_1 + d - \tfrac{1}{2}X_2$ approximately, remembering that each X must have the same sign as its r.

Example. As an example, consider the lens sketched in Fig. 12 having

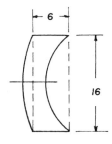

FIG. 12. The volume of a lens.

$r_1 = 20$, $r_2 = 10$, diam. $= 16$, and edge thickness $= 6$. The surface sags are found to be $X_1 = 1.6697$ and $X_2 = 4.00$. The three volumes to be added up are

	Accurate by (11)	Approximate
Convex cap:	54.2π	53.4π
Cylinder:	384.0π	384.0π
Concave cap:	-138.7π	-128.0π
Volume	299.5π	309.4π

The error in the approximate calculation is only 3%, even for a very deeply curved lens such as this.

E. Solution for Last Radius to Give a Stated U'

In some cases we need to determine the last radius of a lens to yield a specified value of the emerging ray slope U', given the Q and U of the incident ray at the surface and the refractive indices n and n'. Since $I' = I + (U - U')$,

$$\sin I' = \sin I \cos(U - U') + \cos I \sin(U - U')$$

and dividing by $\sin I$ gives

$$\sin I'/\sin I = n/n' = \cos(U - U') + \operatorname{ctn} I \sin(U - U')$$

Hence

$$\tan I = \frac{\sin(U - U')}{(n/n') - \cos(U - U')} \tag{12a}$$

Then knowing I we calculate r by

$$r = Q/(\sin U + \sin I) \tag{12b}$$

V. RAY TRACING AT A TILTED SURFACE

So far we have considered only a lens system in which the centers of curvature all lie on a single axis. However, it is sometimes required to consider the effect of a slight tilt of a single surface or element in order to compute a "tilt tolerance" for use in the factory. Special formulas are necessary to trace a meridional ray through such a tilted surface.

A. The Ray Tracing Equations

Suppose the center of curvature of a tilted surface lies at a distance δ to one side of the lens axis. The angular tilt α of the surface is then given by $\sin \alpha = -\delta/r$, the angle α being reckoned positive for a clockwise tilt.

In Fig. 13a, P is the point of incidence of the ray at the tilted surface, C is

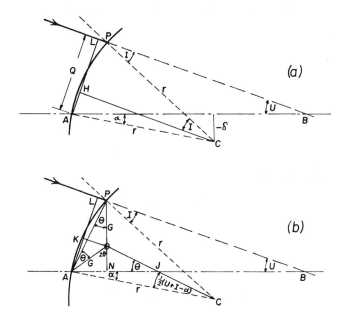

FIG. 13. A ray incident upon a tilted surface.

the center of curvature of the surface distant $-\delta$ below the axis, and angle PCA is clearly equal to $U + I - \alpha$. We draw a line through C parallel to the ray, which intersects the perpendicular AL at H. Thus, Q is equal to $LH + HA$. Angle PCH is equal to I, whence LH is $r \sin I$. The length $HA = r \sin HCA$, where $HCA = PCA - I = U - \alpha$. Consequently,

$$Q = r \sin I + r \sin(U - \alpha) \qquad \text{or} \qquad \sin I = Qc - \sin(U - \alpha)$$

To complete the derivation, we turn to Fig. 13b. Here angle PCA is bisected to intersect the vertical line PN at O. By the congruence of the two triangles POC and AOC, we see that $PO = OA = G$. Angle $APO =$ angle $OJA =$ angle $PAO = \theta$.

However, $\theta = ACJ + JAC = \frac{1}{2}(U + I - \alpha) + \alpha = \frac{1}{2}(U + I + \alpha)$. Therefore, angle $AON = 2 \times APO = (U + I + \alpha)$, whence

$$Y = PN = G[1 + \cos(U + I + \alpha)]$$

$$X = AN = G \sin(U + I + \alpha)$$

To relate Q and G, we draw the usual perpendicular from A onto the ray at L, and draw a line through O parallel to the ray, intersecting Q at the point K. Then

$$Q = LK + KA = G \cos U + G \cos KAO$$

However,

$$KAO = KAN - NAO = (90° - U) - (90° - 2\theta) = 2\theta - U = I + \alpha.$$

Therefore,

$$Q = G[\cos U + \cos(I + \alpha)]$$

or

$$G = Q/[\cos U + \cos(I + \alpha)]$$

The ray tracing equations therefore become

$$\sin I = Qc - \sin(U - \alpha)$$
$$\sin I' = (n/n') \sin I$$
$$U' = U + I - I'$$

Short radius only:

$$Q' = [\sin I' + \sin(U' - \alpha)]/c$$

Universal:

$$G = Q/[\cos U + \cos(I + \alpha)]$$
$$Q' = G[\cos U' + \cos(I' + \alpha)]$$

The transfer to the next surface is normal. In using these equations, it is advisable to list the unusual angles as they arise. They are $U - \alpha$, $I + \alpha$, $I' + \alpha$, and $U + I + \alpha$ for calculating X and Y. It should be noted that a ray running along the axis is refracted at a tilted surface and sets off in an inclined direction, so that paraxial rays have no meaning. An example of ray tracing through a tilted surface is given in Table II. For the calculation of astigmatism through a tilted surface, see p. 192.

A lens element that has been displaced laterally by an amount Δ without otherwise being tilted possesses two tilted surfaces, as indicated in Fig. 14. The tilt of the first surface is $\alpha_1 = \sin^{-1}(-\Delta/r_1)$ and the tilt of the second surface is $\alpha_2 = \sin^{-1}(-\Delta/r_2)$, the Δ being reckoned negative if the lens is displaced below the axis, as in this diagram. Care must be taken to compute the axial separations d along the main axis of the system and not along the displaced axis of the decentered lens element. For small displacements such

TABLE II

TRACING OF THREE RAYS THROUGH A TILTED SURFACE

c	-0.0616427		-0.0616427		-0.0616427	
d	0.4	11.285856	0.4	11.285856	0.4	11.285856
n	1.649	1.0	1.649	1.0	1.649	1.0
	Upper marginal ray		*Axial ray*		*Lower marginal ray*	
$U - \alpha$	2.04276		-1.0		-4.04276	
$U' - \alpha$	7.89180		-1.64914		-11.29834	
Q	1.9186619		0		-1.9186619	
Short radius Q'	1.8900228		0		-1.8715593	
l	-8.85398		1.0		10.88115	
l'	-14.70302		1.64914		18.13673	
$\sin U$	0.0530812	0.1545690	0	-0.0113294	-0.0530812	-0.1787738
U	3.04276	8.89180	0	-0.64914	-3.04276	-10.29834
$l + \alpha$	-7.85398		0		11.88115	
G	0.9645347		0		-0.9704096	
$l' + \alpha$	-13.70302		0.64914		19.13673	
Universal Q'	1.8900237		0		-1.8715603	

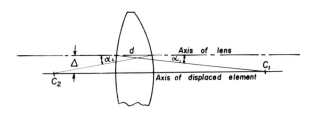

FIG. 14. A decentered lens element.

as might occur by accident this is no problem, but if a lens has been deliberately displaced for some reason, this point must be carefully watched.

B. EXAMPLE OF RAY TRACING THROUGH A TILTED SURFACE

We will take the cemented doublet lens of p. 28 and imagine that the rear surface has been tilted clockwise by $\alpha = 1°$. We shall have to trace the axial ray, the upper marginal ray, and the lower marginal ray, because all three of these rays are treated differently by a tilted surface (see Table II).

To understand what has happened as a result of tilting the rear surface by $1°$, we calculate the height at which each emerging ray crosses the paraxial focal plane:

$$\begin{array}{rl}
\text{upper marginal ray:} & 0.147350 \\
\text{axial ray:} & 0.127870 \\
\text{lower marginal ray:} & 0.148448
\end{array}$$

In Fig. 15 we have plotted on a large scale this situation as compared with

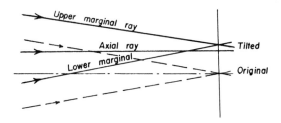

FIG. 15. The result of tilting a lens surface.

the case before the surface was tilted. It is clear that the entire image has been raised, and there is a large amount of coma introduced by the tilting.

VI. RAY TRACING AT AN ASPHERIC SURFACE

An aspheric surface can be defined in several ways, the simplest being to express the sag of the surface from a plane:

$$X = a_2 Y^2 + a_4 Y^4 + a_6 Y^6 + \cdots$$

Only even powers of Y appear because of the axial symmetry. The first term is all that is required for a parabolic surface. To express a sphere in this way we use the power series given in Eq. (9), but a great many terms will be required if the sphere is at all deep.

For many purposes it is better to express the asphere as a departure from a sphere:

$$X = \frac{cY^2}{1 + (1 - c^2 Y^2)^{1/2}} + a_4 Y^4 + a_6 Y^6 + \cdots \tag{13}$$

Here c represents the curvature of the osculating sphere and $a_4, a_6, \ldots$ are the aspheric coefficients.

If the surface is known to be a conic section, we may express it by

$$X = \frac{cY^2}{1 + [1 - c^2 Y^2 (1 - e^2)]^{1/2}} \tag{14}$$

where c is the vertex curvature of the conic and e its eccentricity. The term $1 - e^2$ in this expression is called the "conic constant" since it defines the shape of the surface. Its value is as follows:

Surface	Eccentricity	Conic constant
Hyperbola	> 1	negative
Parabola	1	0
Prolate spheroid (small end of ellipse)	< 1	< 1
Sphere	0	1
Oblate spheroid (side of ellipse)	—	> 1

To trace a ray through an aspheric surface, we must first determine the X and Y coordinates of the point of incidence. The asphere is defined by a relation between X and Y, while the incident ray is defined by its Q and U. Now it is clear from Fig. 16 that

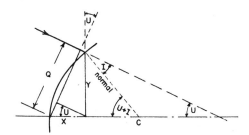

FIG. 16. Ray trace through an aspheric surface.

$$Q = [X] \sin U + Y \cos U$$

where $[X]$ is to be replaced by the expression for the aspheric surface, giving an equation for Y having the same order as the asphere itself.

To solve this equation, we first guess a possible value of Y, say, $Y = Q$. We then evaluate the residual R as follows:

$$R = [X] \sin U + Y \cos U - Q$$

Obviously the correct value of Y is that which makes $R = 0$. Now Newton's rule says that

$$(\text{a better } Y) = (\text{the original } Y) - (R/R')$$

where R' is the derivative of R with respect to Y, namely,

$$R' = (dX/dY) \sin U + \cos U$$

A very few iterations of this formula will give us the value of Y that will make R less than any defined limit, such as 0.00000001. Knowing Y we immediately find X from the equation of the asphere. We then proceed as follows:

The slope of the normal is dX/dY. Hence

$$\tan(U + I) = dX/dY$$

$$\sin I' = (n/n') \sin I$$

$$U' = U + I - I'$$

$$Q' = X \sin U' + Y \cos U'$$

The transfer to the next surface is standard.

This process can be simplified in the case of a surface that is a conic section, because the equation to be solved is then an ordinary quadratic. Note that if the asphere is defined by Eq. (14) the derivative becomes

$$\tan(U + I) = dX/dY = cY/[1 - c^2 Y^2 (1 - e^2)]^{1/2} \tag{15}$$

Example. Suppose our asphere is given by

$$[X] = 0.1 Y^2 + 0.01 Y^4 - 0.001 Y^6$$

Then

$$dX/dY = 0.2 Y + 0.04 Y^3 - 0.006 Y^5$$

with $n = 1.0$ and $n' = 1.523$. If our entering ray has $U = -10°$ and $Q = 3.0$, then successive iterations of Newton's rule give

	Y	X	dX/dY	R	R'	R/R'
1.	3.0	0.981	0.222	0.124772	1.023358	0.121924
2.	2.878076	0.946119	0.344369	−0.001357	1.044607	−0.001299
3.	2.879375	0.946566	0.343244	0		

Hence

$$\tan(U + I) = (dX/dY) = 0.343244, \qquad U + I = 18.94448°$$

But $U = -10°$. Therefore

$$I = 28.94448°, \qquad I' = 18.52805°$$

$$U' = 0.41644$$

$$Q' = Y \cos U' + X \sin U' = 2.886178$$

CHAPTER 3

Paraxial Rays and First-Order Optics

Suppose we trace a number of meridional rays through a lens from a given object point, the incidence heights varying from the marginal ray height Y_m down to a ray lying very close to the lens axis. We then plot a graph (Fig. 17) connecting the incidence height Y with the image distance L'. This graph will

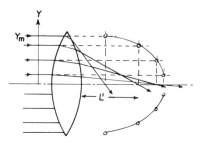

FIG. 17. Plot of Y against L'.

have two branches, the half below the axis being identical with that above the axis but inverted. The precision of the various point locations is good at the margin but drops badly when the ray is very close to the lens axis, and actually at the axis there is no precision at all. Thus by ordinary ray tracing we can plot all of this graph with the exception of the portion lying near the axis, and we cannot in any way find the exact point at which the graph actually crosses the axis. This failure is, of course, due to the limited precision of our mathematical tables and our computing procedures.

However, the exact point at which the graph crosses the axis can be found as a limit. A ray lying everywhere very close to the optical axis is called a "paraxial" ray, and we can regard the paraxial image distance l' as the limit toward which the true L' tends as the aperture Y is made progressively smaller, or

$$l' = \lim_{y \to 0} L'$$

I. TRACING A PARAXIAL RAY

Since all paraxial heights and angles are infinitesimal, we can determine their relative magnitudes by use of a new set of ray-tracing equations formed by writing sines equal to the angles in radians, and cosines equal to 1.0. Since infinitesimals have finite relative magnitudes, we may use any finite numbers to represent paraxial quantities, but we must remember to assume that each number is to be multiplied by a very small factor such as 10^{-50}, so that a paraxial angle written 2.156878 does not mean 2.156878 rad but 2.156878×10^{-50} rad. It is quite unnecessary to write the 10^{-50} every time, but its existence must be assumed if paraxial quantities are to have any meaning. Of course, the longitudinal paraxial data such as l and l' are not infinitesimals.

A. THE STANDARD PARAXIAL RAY TRACE

Once this is understood, we can derive a set of equations for tracing paraxial rays by modifying Eqs. (4). Writing sines as angles and cosines as unity, and remembering that in the paraxial region both Q and Q' degenerate to the paraxial ray height y, we get

$$i = yc - u, \qquad y = lu = l'u'$$
$$i' = (n/n')i \qquad\qquad\qquad (16)$$
$$u' = u + i - i' = yc - i'$$

with the transfer $y_2 = y_1 - du'_1$. It will be noticed that paraxial quantities are written with lowercase letters to distinguish them from true heights and angles such as are used in computing the path of a real ray.

As an example, we will repeat the lens data given on p. 28 for a cemented doublet, and trace a paraxial ray through it with the starting data $y = 2.0$ and $u = 0$ (Table III). As before, the paraxial image distance is found by dividing the last y by the emerging u', giving $l' = 11.285849$. This is slightly different from the marginal L, which was found to be 11.293900. The difference is caused by spherical aberration.

B. THE $(y - nu)$ METHOD

Because of the linear nature of paraxial relationships, we can readily submit the paraxial ray-tracing equations to algebraic manipulation to eliminate some or all of the paraxial angles, which are actually only auxiliary quantities. For example, to eliminate the angles of incidence i and i', we

TABLE III

TRACING A PARAXIAL RAY THROUGH A CEMENTED DOUBLET

c	0.1353271		−0.1931098		−0.0616427	
d		1.05		0.4		
n		1.517		1.649		
y	2		1.9031479		1.8809730	
i	0.2706542		−0.4597566		−0.1713856	
i'	0.1784141		−0.4229538		−0.2826148	
u	0	0.0922401		0.0554373		0.1666665

multiply the first of Eq. (16) by n and the corresponding expression for the refracted ray by n', giving

$$ni = nyc - nu, \qquad n'i' = n'yc - n'u'$$

Now the law of refraction for paraxial rays is merely $ni = n'i'$; hence equating these two expressions gives

$$n'u' = nu + y(n' - n)c \qquad (17a)$$

This formula can be used to trace paraxial rays, in conjunction with the transfer

$$y_2 = y_1 + (-t/n)(n'_1 u'_1) \qquad (17b)$$

It will be noticed that, written in this way, Eqs. (17a) and (17b) are of the same form. That is, in each equation the new value is found by taking the former value and adding to it the product of the other variable multiplied by a constant. This leads to a remarkably convenient and simple ray-tracing procedure known as the $(y - nu)$ method. In Table IV we have traced the paraxial ray of Table III by these equations.

The operating procedure is as follows. To calculate each number, be it a y or an nu, we take the previous y or nu and add to it the product of the next number to the right multiplied by the constant located immediately above it. Thus, starting with y_1 and $(nu)_1$, we first find $(nu)'_1 = (nu)_1 + y_1(n' - n)_1 c_1$. Then for y_2 we take y_1 and add to it the product of $(nu)'_1$ and $-d/n$, and so on in a zigzag manner right through to the last surface. The closing equation is, of course,

$$l' = (\text{last } y)/[\text{last } (nu)']$$

The numbers in the $(y - nu)$ ray-tracing table obviously resemble perfectly the corresponding numbers in Table III, where the paraxial ray was traced by conventional means. The amount of work involved in the $(y - nu)$

TABLE IV

THE $(y - nu)$ METHOD FOR TRACING PARAXIAL RAYS

c	0.1353271		0.1931098		−0.0616427	
d		1.05		0.4		
n		1.517		1.649		
$\phi = (n' - n)c$	0.0699641		−0.0254905		0.0400061	
$-d/n$		−0.6921556		−0.2425713		
y	2		1.9031479		1.8809730	$l' = 11.285856$
nu	0		0.1399282		0.0914160	0.1666664
l	∞		20.632549		33.929774	
l'	21.682549		34.329774		11.285856	

method is about the same as by the direct method, but there are many advantages in tracing rays this way, as we shall see.

Since the image distance l' is the same for all paraxial rays starting out from the same object point, we may select any value we please for either the starting y or the starting nu, but not both, since they are related by $y = lu$. Many designers always use $y_1 = 1.0$ and calculate the appropriate value of $(nu)_1$. Thus if an object is located at 50 units to the left of the first surface, we could take $y_1 = 1.0$ and $(nu)_1 = -0.02$, remembering that l is negative if the object lies to the left of the surface. A positive l implies a virtual object lying to the right of the first lens surface when the entering rays come in from the left.

When tracing a paraxial ray backwards from right to left, we must subtract each product from the previous value instead of adding it. Thus for right-to-left work we have

$$nu = (nu)' - y(n' - n)c, \qquad y_1 = y_2 - (-d/n)(nu)_2$$

C. INVERSE PROCEDURE

One advantage of the $(y - nu)$ method over the straightforward procedure using i and i' is that we can, if we wish, invert the process and work upward from the ray data to the lens data. Thus if we know from some other considerations the succession of y and nu values, we can calculate the lens data by inverting Eqs. (17a) and (17b) giving

$$\phi = \frac{n' - n}{r} = \frac{(nu)' - nu}{y} \qquad \frac{-d}{n} = \frac{y_2 - y_1}{nu}$$

This is often an extremely useful procedure, which cannot be performed when using the straightforward ray trace.

D. ANGLE SOLVE AND HEIGHT SOLVE METHODS

When making changes in a lens, it is sometimes desired to maintain either the height of incidence of a paraxial ray at a particular surface by a change in the preceding thickness, or to maintain the paraxial ray slope after refraction by a change in the curvature of the surface. Both of these can be achieved by an inversion of Eqs. (17a) and (17b). Thus for a height solve we determine the prior surface separation by

$$d = (y_1 - y_2)/u'_1$$

and for an angle solve we use

$$c = [(nu)' - nu]/y(n' - n)$$

The last formula is particularly useful if we wish to maintain the focal length of a lens by a suitable choice of the last radius. It should be noted that this formula is the paraxial equivalent of Eq. (12), obtained by writing i for tan I, $(u - u')$ for $\sin(U - U')$, 1.0 for $\cos(U - U')$, u and i for sin U and sin I, respectively, and y for Q.

E. THE (l, l') METHOD

In the derivation of Eq. (16) we eliminated the angles of incidence as being unnecessary auxiliaries. Actually we can go further and also eliminate the ray slope angles u and u'. To do this we divide Eq. (16) by y and note that $l = y/u$, while $l' = y/u'$. These substitutions give the well-known expression

$$\frac{n'}{l'} = \frac{n}{l} + \frac{n' - n}{r} \tag{18}$$

In computations, this is used in the form

$$l' = \frac{n'}{(n/l) + \phi}$$

where

$$\phi = (n' - n)/r = (n' - n)c$$

The transfer now is merely

$$l_2 = l'_1 - d$$

An example of the results of this calculation is given at the bottom of Table IV. Remember that l and l' refer to the portions of a ray lying to left and right of a surface, respectively. Of course, in the spaces between surfaces the ray almost never reaches the optical axis, so that neither the l nor the l' is actually realized.

F. A Paraxial Ray at an Aspheric Surface

In tracing a paraxial ray, the aspheric terms have no effect and we need to consider only the vertex curvature of the surface. This is given by the coefficient of the second-order term in the power series expansion.

G. Graphical Tracing of Paraxial Rays

We cannot use the ordinary graphical ray-tracing process for paraxial rays since they fall too close to the axis to be traced. However, we can imagine the entire set of curves and lines to be drawn on sheet rubber and then stretched transversely by 100 or more times. When this is done all the circles in the ordinary process become straight lines transverse to the lens axis, but all points lying in the axis retain their positions, including the center of curvature of the surface and the paraxial object and image points. The paraxial equivalent of Fig. 8 appears in Fig. 18. The lateral scale is

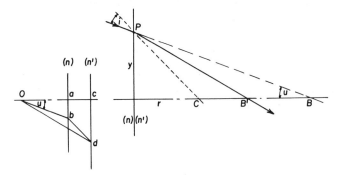

Fig. 18. Graphical tracing of a paraxial ray.

unimportant. The heights of incidence are the same as the computed y values; the length ab is equal to nu and the length cd is equal to $(nu)'$. Since ac is $n' - n$, it is clear that $(nu)' = nu + y(n' - n)/r$ in accordance with Eq. (17a).

II. RAY TRACING BY THE (W, U) METHOD

A greatly simplified method is available[1] for tracing both real and parax-

[1] M. Herzberger, "Modern Geometrical Optics," p. 18. Wiley (Interscience), New York, 1958.

ial rays, provided the radius of curvature is short or plane. The ray to be traced is defined by its W and its U, where W is defined as the length of the perpendicular to the ray drawn from the center of curvature of the surface multiplied by the refractive index of the medium in which the ray lies. Thus (Fig. 19a)

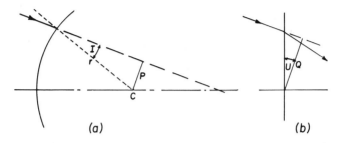

FIG. 19. (a) Tracing showing that $P = r \sin I$. (b) A plane surface.

$$W = nP = nr \sin I = n'r \sin I' \text{ (by the law of refraction)} = n'P'$$

In this scheme the ray-tracing equations are

	Marginal		*Paraxial*
$\sin I = WR$	(where $R = c/n$)	$i = wR$	
$\sin I' = WR'$	(where $R' = c/n'$)	$i' = wR'$	
$U' = U + I - I'$		$u' = u + i - i' = u + wZ$	
		[where $Z = (R - R')$]	

Opening:

$$W = n(L - r) \sin U$$

Transfer:

$$W_2 = W_1 + C \sin U'_1 \quad \text{where} \quad C = n(r_1 - d - r_2)$$

Closing:

$$L' = W/(n' \sin U') + r$$

An example of the tracing of a marginal ray and a paraxial ray by this method is shown in Table V. Incidentally, the paraxial quantity Z is nothing but the Petzval term $(n' - n)/nn'r$. The set-up for this method is more complex than for the previous methods, but the ray tracing is much simpler; hence we tend to use this method only if we are expecting to trace several rays through the same lens. We cannot use this method for height solve or angle solve operations.

TABLE V

RAY TRACING BY THE (W, U) METHOD

						Marginal / Paraxial						
r	8.572		-7.258		∞		5.735		-3.807		-16.878	
d		2.4		0.4		7.738		1.8		0.4		
n		1.5224		1.61644		1		1.51625		1.61644		
R	0.1166589		-0.0905012				0.1743679		-0.1732393		-0.0366538	
R'	0.0766283		-0.0852361				0.1149994		-0.1625016		-0.0592487	
Z	0.0400306		-0.0052651				0.0593685		-0.0107377		0.0225949	
C		20.44583		-12.3787		-13.473		11.73881		20.48191		

Marginal ray

W	3.175		5.899857		$\begin{cases}4.696820\\2.883227\end{cases}$		0.7666809		3.139677		6.481176	
I	21.73980		-32.27233				7.68256		-32.95058		-13.74256	
I'	14.08109		-39.19077				5.05821		-30.67732		-22.58175	
U	0	7.65871		5.57715		9.03834	11.66269		9.38943		18.22862	

Paraxial ray

w	3.175		5.773607		$\begin{cases}4.576604\\2.831286\end{cases}$		0.7253551		3.065730		6.474981	
u	0	0.1270972		0.0966986		0.1563075	0.1993707		0.1664518		0.3127533	

Closing: $L' = 3.841251$ $f' = 10.151771$
$l' = 3.825158$
$LA' = \overline{0.016093}$

When using this procedure, it is useful to store the value of W in the calculator memory since it is used three times at each surface. It is not necessary to record the sines of the angles since they are not required again.

A. SPECIAL CASE OF A PLANE SURFACE

Obviously the ordinary (W, U) formulas are inapplicable to a plane surface since the center of curvature is at infinity and W would become infinite. However, we can use a modified form of the standard (Q, Q') procedure by expressing W as nQ and W' as $n'Q'$. Then for a plane surface we have (Fig. 19b)

$$\sin U' = (n/n') \sin U, \qquad u' = nu/n'$$

$$W' = W \tan U/\tan U', \qquad w' = wn'/n$$

The transfer term has also lost its meaning since one of the radii is now infinite. However, we can transfer to or from a plane surface by merely letting the plane surface count as $r = 0$ when calculating C for the transfer.

Also, for a plane we have always $X = 0$. Then $Y = W/(n \cos U) = W'/(n' \cos U')$. If the last surface is a plane, the closing equation becomes $L = W'/(n' \sin U')$.

III. MAGNIFICATION AND THE LAGRANGE THEOREM

A. TRANSVERSE MAGNIFICATION

Consider first a single refracting surface as in Fig. 20. Let B, B' be a pair

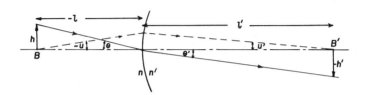

FIG. 20. The Lagrange relationship.

of axial conjugate points, their distances from the surface being l and l', respectively. We now place a small object at B and draw a paraxial ray from the top of the object to the vertex of the surface. The ray will be refracted there, the slope angles θ and θ' being the angles of incidence and refraction. Hence $n\theta = n'\theta'$, and therefore $nh/l = n'h'/l'$. Multiplying both sides by y gives

$$hnu = h'n'u' \qquad (19)$$

This important relationship is called the theorem of Lagrange, or sometimes the Smith–Helmholtz theorem. Because the h', n', and u' on the right of one surface are respectively equal to the same quantities on the left of the next surface, it is clear that the product hnu is invariant for all the spaces between surfaces, including the object space and the image space. This product is called the Lagrange invariant or, more often nowadays, the optical invariant.

Since this theorem applies to the original object and also the final image, it is clear that the image magnification is given by

$$m = h'/h = nu/n'u'$$

For a lens in air, the magnification is merely u_1/u'_k (assuming that there are k surfaces in the system). The fact that the ratio of the nu values at the object and at the image determines the magnification is one of the reasons why it is usually preferred to trace a paraxial ray by the $(y - nu)$ method.

B. Longitudinal Magnification

If an object has a small longitudinal dimension δl along the lens axis, or if it is moved along the axis through a small distance δl, then the corresponding longitudinal image dimension is $\delta l'$, and the longitudinal magnification $\bar{m}$ is given by $\bar{m} = \delta l'/\delta l$. By differentiating Eq. (18) we find

$$-n'\,\delta l'/l'^2 = -n\,\delta l/l^2$$

and multiplying both sides by y^2 gives

$$n'\,\delta l'u'^2 = n\,\delta l u^2$$

This is the longitudinal equivalent of the Lagrange equation, and the product $n\,\delta l u^2$ is also an invariant. The longitudinal magnification is found to be

$$\bar{m} = \delta l'/\delta l = nu^2/n'u'^2 = (n'/n)m^2 \tag{20}$$

so that for a lens in air, $\bar{m} = m^2$. Hence longitudinal magnification is always positive, meaning that if the object is moved a short distance from left to right the image will move from left to right also.

IV. THE GAUSSIAN OPTICS OF A LENS SYSTEM

It is generally considered that Gauss was responsible for the concept of the four cardinal points and two focal lengths of a lens. To understand the nature of these terms, we imagine a family of parallel rays entering the lens from the left in a direction parallel to the axis (Fig. 21). A marginal ray such

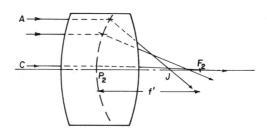

FIG. 21. The equivalent refracting locus.

as A will, after passing through the lens, cross the axis in the image space at J, and so on down to the paraxial ray C, which crosses the axis finally at F_2.

If the entering and emerging portions of all these rays are extended until they intersect, we can construct an "equivalent refracting locus" as a surface of revolution about the lens axis, to contain all the equivalent refracting points for the entire parallel beam. The paraxial portion of this locus is a plane perpendicular to the axis and known as the *principal plane*, and the axial point itself is called the *principal point*, P_2. The paraxial image point F_2, which is conjugate to infinity, is called the *focal point*, and the longitudinal distance from P_2 to F_2 is the posterior *focal length* of the lens, marked f'.

A beam of parallel light entering parallel to the axis from the right will similarly yield another equivalent refracting locus with its own principal point P_1 and its own focal point F_1, the separation from P_1 to F_1 being known as the anterior focal length f. The distance from the rear lens vertex to the F_2 point is the *back focal distance* or more commonly the *back focus* of the lens, and of course the distance from the front lens vertex to the F_1 point is the *front focus* of the lens. For historical reasons the focal length of a compound lens has often been called the *equivalent focal length*, or EFL, but the term equivalent is redundant and it will not be used here.

A. The Relation between the Principal Planes

Proceeding further, we see in Fig. 22 that a paraxial ray A traveling from

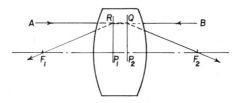

FIG. 22. The principal planes as unit planes.

left to right is effectively bent at the second principal plane Q and emerges through F_2, while a similar paraxial ray B traveling from right to left along the same straight line will be effectively bent at R and cross the axis at F_1. Reversing the direction of the arrows along ray BRF_1 yields two paraxial rays entering from the left toward R, which become two paraxial rays leaving from the point Q to the right; thus Q is obviously an image of R, and the two principal planes are therefore conjugates. Because R and Q are at the same height above the axis, the magnification is $+1$, and for this reason the principal planes are sometimes referred to as "unit planes."

When any arbitrary paraxial ray enters a lens from the left it is continued until it strikes the P_1 plane, and then it jumps across the "hiatus" between the principal planes, leaving the lens from a point on the second principal plane at the same height as it encountered the first principal plane (see Fig. 23).

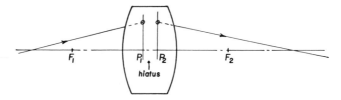

FIG. 23. A general paraxial ray traversing a lens.

B. RELATION BETWEEN THE TWO FOCAL LENGTHS

Suppose a small object of height h is located at the front focal plane F_1 of a lens (Fig. 24). We draw a paraxial ray parallel to the axis from the top of

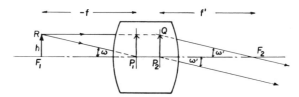

FIG. 24. Ratio of the two focal lengths.

this object into the lens; it will be effectively bent at Q and emerge through F_2 at a slope ω'. A second ray from R directed toward the first principal point P_1 will emerge from P_2 because P_1 and P_2 are images of each other, and it will emerge at the slope ω' because R is in the focal plane and therefore all rays starting from R must emerge parallel to each other on the right-hand side of the lens. From the geometry of the figure, $\omega = -h/f$ and $\omega' = h/f'$; hence

$$\omega'/\omega = -f/f' \tag{21a}$$

We now move the object h along the axis to the first principal plane P_1. Its image will have the same height and will be located at P_2. We can now apply the Lagrange theorem to this object and image, knowing that a paraxial ray is entering P_1 at slope ω and leaving P_2 at slope ω'. Therefore, by Lagrange,

$$hn\omega = hn'\omega' \quad \text{or} \quad \omega'/\omega = n/n' \tag{21b}$$

Equating (21a) and (21b) tells us that

$$f/f' = -n/n'$$

IV. THE GAUSSIAN OPTICS OF A LENS SYSTEM

The two focal lengths of any lens, therefore, are in proportion to the outside refractive indices of the object and image spaces. For a lens in air, $n = n' = 1$, and the two focal lengths are equal but of opposite sign. This negative sign simply means that if F_1 is to the right of P_1 then F_2 must lie to the left of P_2. It does *not* mean that the lens is a positive lens when used one way round and a negative lens when used the other way round. The sign of the lens is the same as the sign of its posterior focal length f'. For a lens used in an underwater housing, $n = 1.33$ and $n' = 1.0$; hence the anterior focal length is 1.33 times as long as the posterior focal length.

C. LENS POWER

Lens power is defined as

$$P = \frac{n'}{f'} = -\frac{n}{f}$$

Thus for a lens in air the power is the reciprocal of the posterior focal length. Focal length and power can be expressed in any units, of course, but if focal length is given in meters, then power is in diopters. Note too that the power of a lens is the same on both sides no matter what the outside refractive indices may be.

Applying Eq. (17a) to all the surfaces in the system and summing, we get

$$\text{power} = P = \frac{(nu')_k}{y_1} = \sum \frac{y}{y_1}\left(\frac{n' - n}{r}\right) \tag{22}$$

The quantity under the summation is the contribution of each surface to the lens power. The expression in parentheses, namely, $(n' - n)/r$, is the *power* of a surface.

D. CALCULATION OF FOCAL LENGTH

1. *By an Axial Ray*

If a paraxial ray enters a lens parallel to the axis from the left at an incidence height y_1 and emerges to the right at a slope u' (see Fig. 25a), then the posterior focal length is $f' = y_1/u'$. The anterior focal length f is found similarly by tracing a parallel paraxial ray right to left, and of course we find that $f = -f'$ if the lens is in air. The distance from the rear lens vertex to the second principal plane is given by

$$l'_{pp} = l' - f'$$

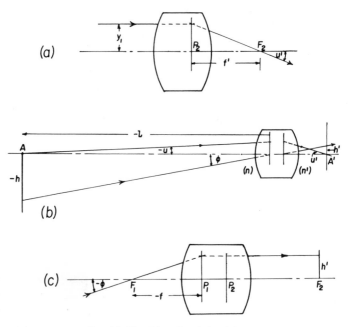

FIG. 25. Focal-length relationships.

and similarly

$$l_{pp} = l - f$$

2. By an Oblique Ray

The Lagrange equation can be modified for use with a very distant object in the following way. In Fig. 25b, let A represent a very distant object and A' its image. As the object distance l becomes infinite, the image A' approaches the rear focal point. Then by Lagrange's theorem

$$h'n'u' = hnu = (h/l)n(lu) = -ny_1 \tan \phi$$

or

$$h' = -\tan \phi \cdot (n/n') \cdot f' = +f \tan \phi \tag{23}$$

where f is the anterior focal length of the lens, no matter what the outside refractive indices may be. This equation forms the basis of the current ANSI definition of focal length. Actually this relation is obvious from a consideration of Fig. 25c, where a paraxial ray is shown entering a lens through the anterior focal point at a slope angle ϕ.

E. CONJUGATE DISTANCE RELATIONSHIPS

It is easy to show by similar triangles that if the distances of object and image from the corresponding focal points of a lens are x and x', then

$$m = -f/x = -x'/f' \quad \text{whence} \quad xx' = ff' \tag{24a}$$

Similarly, if the distances of object and image from their respective principal points are p and p', then

$$n'/p' - n/p = n'/f' = -n/f = \text{lens power} \quad \text{and} \quad m = np'/n'p \tag{24b}$$

For a lens in air this becomes simply

$$\frac{1}{p'} - \frac{1}{p} = \frac{1}{f'} \quad \text{and} \quad m = \frac{p'}{p} \tag{24c}$$

It is often convenient to combine the last two equations for the usual case of a positive lens forming a real image of a real object. Furthermore, if we then ignore all signs and regard all dimensions as positive, with a positive magnification, we get

$$f' = \frac{pp'}{p + p'}, \quad p = f'\left(1 + \frac{1}{m}\right), \quad p' = f'(1 + m) \tag{25a}$$

These relations are often expressed verbally as "Object distance is $[1 + (1/m)]$ focal lengths, and image distance is $(1 + m)$ focal lengths."

Combining these we get an expression for the object-to-image distance D as

$$D = f'\left(2 + m + \frac{1}{m}\right) \tag{25b}$$

Inverting this we can calculate the magnification when we are given f' and D:

$$m = \tfrac{1}{2}k - 1 \pm (\tfrac{1}{4}k^2 - k)^{1/2} \quad \text{where} \quad k = D/f' \tag{25c}$$

It is important to understand that p and x refer to that section of the ray that lies to the left of the lens, no matter whether that ray actually crosses the axis to the left of the lens, and no matter whether that ray defines the "object" or the "image" in any particular situation. Similarly, p' and x' refer to the section of a ray lying to the right of the lens. The p' and x' are positive if they lie to the right of their origins, namely, the second principal point and the second focal point, respectively.

F. Nodal Points

The nodal points of a lens are a pair of conjugate points on the lens axis such that any paraxial ray directed toward the first nodal point emerges from the second at the same slope at which it entered. If the nodal points are N_1 and N_2 and the principal points are P_1 and P_2, then it is easy to show that (Fig. 26)

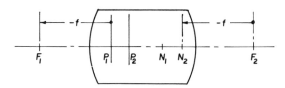

FIG. 26. The principal and nodal points.

$$F_1 P_1 = F_2 N_2 = f \quad \text{and} \quad F_1 N_1 = P_2 F_2 = f'$$

If the lens is in air, the two focal lengths are equal, and the nodal points coincide respectively with the two principal points.

V. FIRST-ORDER LAYOUT OF AN OPTICAL SYSTEM

Most optical systems, as opposed to a specific objective lens, are assembled first from a series of "thin" lens elements at finite separations, and it is therefore of interest to collect here a few useful relations governing the properties of a single thick lens and a set of thin lenses.

A. A Single Thick Lens

By setting up the familiar $(y - nu)$ table for the two surfaces of a single thick lens, it is easy to show that

$$\text{power} = \frac{1}{f'} = (N-1)\left(\frac{1}{r_1} - \frac{1}{r_2} + \frac{t}{N} \frac{N-1}{r_1 r_2} \right)$$

where N is the refractive index of the glass. The back focus is given by

$$l' = f'\left(1 - \frac{t}{N} \frac{N-1}{r_1} \right)$$

and the rear principal plane is located at

$$l'_{pp} = l' - f' = -f'\left(\frac{t}{N} \frac{N-1}{r_1} \right)$$

Similar relations exist for the front focal length and front focal distance. The hiatus or separation between the two principal planes is

$$Z = t + l'_{pp} - l_{pp} = t(N - 1)/N$$

For common crown glass with a refractive index of approximately 1.5, the value of Z is about $t/3$.

B. A SINGLE THIN LENS

If a lens is so thin that, within the precision in which we are interested, we can ignore the thickness for any calculations, then we can regard it as a "thin" lens. For accurate work, of course, no lens is thin. Nevertheless, the concept of a thin lens is so convenient in the preliminary layout of optical systems that we often use thin-lens formulas in the early stages of a design and insert thickness for the final studies.

The power of a thin lens is the sum of the powers of its component surfaces, or component elements if it is a multielement thin system. This is because an entering ray remains at the same height y throughout the thin system. Hence for a single lens,

$$\text{power} = \frac{1}{f'} = (N - 1)\left(\frac{1}{r_1} - \frac{1}{r_2}\right)$$

and for a thin system,

$$\text{power} = \sum 1/f$$

C. A MONOCENTRIC LENS

A lens in which all the surfaces are concentric about a single point is called *monocentric*. The nodal points of such a lens are, of course, at the common center because any ray directed toward this center is undeviated. Hence the principal and nodal points also coincide at the common center. The image of a distant object is also a sphere centered about the same common center, of radius equal to the focal length. Monocentric systems can be entirely refracting or may include reflecting surfaces.

D. IMAGE SHIFT CAUSED BY A PARALLEL PLATE

It is easy to show (see p. 119) that if a parallel plate of transparent material is inserted between a lens and its image, the image will be displaced further from the lens by an amount

$$s = t\left(1 - \frac{1}{N}\right)$$

Thus, if $N = 1.5$, s will be one-third the thickness of the plate. The image magnification is unity, and this is a well-known method for displacing an image longitudinally without altering its size. A prism lying between a lens and its image also displaces the image by this distance measured along the ray path in the prism; however, the actual physical image displacement will depend on the folding of the ray path inside the prism, and it is possible to devise such a prism that it may be inserted or removed without any physical shift of the final image.

E. LENS BENDING

One of the most powerful tools available to the lens designer is "bending," i.e., changing the shape of an element without changing its power. If the lens is thin, we know that its focal length is given by

$$\frac{1}{f'} = (N - 1)\left(\frac{1}{r_1} - \frac{1}{r_2}\right)$$

We may write $c_1 = 1/r_1$ and $c_2 = 1/r_2$. Then $c = c_1 - c_2$ and we have

$$1/f' = (N - 1)(c_1 - c_2) = (N - 1)c$$

So long as we retain the value of c, we can obviously select any value of c_1 and solve for c_2. If our thin system contains several thin elements, we can state c_1 and then find the other radii by

$$c_1 \text{ given:} \qquad \text{then} \quad c_2 = c_1 - c_a, \quad c_3 = c_2 - c_b, \quad \text{etc.}$$

Alternatively, we can take the data of a given lens and change each surface curvature by the same amount Δc. Then

$$\text{new } c_1 = \text{old } c_1 + \Delta c$$

$$\text{new } c_2 = \text{old } c_2 + \Delta c$$

$$\vdots$$

In Fig. 27 is shown a series of bendings of a lens in which $c = 0.2$, and we start with a bending having $c_1 = -0.1$. We then add $\Delta c = 0.1$ each time, giving the set of lens shapes shown here. Note that a positive bending bends the top and bottom of the lens to the right, whereas a negative bending turns them to the left.

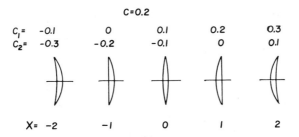

FIG. 27. Bending a single thin lens.

A convenient dimensionless shape parameter X has been used to express the shape of a single lens. It is defined by

$$X = \frac{r_2 + r_1}{r_2 - r_1} = \frac{c_1 + c_2}{c_1 - c_2}$$

Then if we are given f' and X, we can solve for the surface curvatures of a thin lens by

$$c_1 = \tfrac{1}{2}c(X + 1) \qquad \text{and} \qquad c_2 = \tfrac{1}{2}c(X - 1)$$

or

$$c_1 = \frac{X + 1}{2f'(N - 1)} \qquad \text{and} \qquad c_2 = \frac{X - 1}{2f'(N - 1)}$$

Note that for an equiconvex or equiconcave lens $X = 0$. A plano lens has an X value of $+1.0$ or -1.0, while X values greater than 1.0 indicate a meniscus element. X is always positive when the lens is bent to the right and negative to the left.

If the lens to be bent is thick, and especially if it is compound, we can bend it by applying the same Δc to all the surfaces except the last, and then solve the last radius to give the desired lens power by holding the final u'. This is an angle solve problem, discussed on p. 43. However, if the lens is a single thick element, we can still use the X notation for the lens shape if we wish. For a thick lens of focal length f', we find that

$$r_1 = (N - 1)\frac{f' \pm [f'^2 + (f't/N)(X + 1)(X - 1)]^{1/2}}{X + 1}$$

or

$$c_1 = \frac{-N \pm [N^2 + (Nt/f')(X + 1)(X - 1)]^{1/2}}{t(N - 1)(X + 1)}$$

We can then find r_2 or c_2 by the relation

$$r_2 = r_1\left(\frac{X+1}{X-1}\right) \quad \text{or} \quad c_2 = c_1\left(\frac{X-1}{X+1}\right)$$

Example. If $f' = 8$, $t = 0.8$, $N = 1.523$, and $X = 1.2$, then the thin-lens formulas give $c_1 = 0.26291$ and $c_2 = 0.02390$. If the thickness is taken into account, the thick-lens formulas give $c_1 = 0.26103$ and $c_2 = 0.02373$. The effect of the finite thickness is remarkably small, even for a meniscus lens such as this.

F. A SERIES OF SEPARATED THIN ELEMENTS

In the case of a series of separated thin elements we cannot merely add the lens powers to get the power of the system because the y at each element varies with the separations. Instead we must use the result of Eq. (22), namely,

$$\text{power} = \sum (y/y_1)\phi$$

where ϕ is the power of each element.

The familiar $(y - nu)$ ray-tracing procedure can be conveniently applied to a series of separated thin lenses of power ϕ and separation d, noting that the refractive indices appearing in the $(y - nu)$ method are now all unity. The equations to be used are

$$u' = u + y\phi \quad \text{and} \quad y_2 = y_1 + (-d_1')u_1' \tag{26}$$

As an example, we will determine the power and image distance of the following system:

$$\phi_a = 0.125, \quad \phi_b = -0.20, \quad \phi_c = 0.14286$$

$$d_a' = 2.0, \quad d_b' = 3.0$$

The $(y - u)$ table for this system becomes

ϕ	0.125		-0.2		0.14286	
$-d$		-2.0		-3.0		
y	1		0.75		0.825	
u	0	0.125		-0.025		0.09286

Hence the focal length is $1/0.09286 = 10.769$, and the back focus is $0.825/0.09286 = 8.885$. Of course, as always, the (y, u) process is reversible, and if we know what values of y and u the ray should have, we can readily work upwards in the table and find what lens system will give the desired ray path.

As an example, suppose we have two lenses at 2-in. intervals between a fixed object and image 6 in. apart, and we wish to obtain a magnification of −3 times. What must be the powers of the two lenses? We proceed to fill out

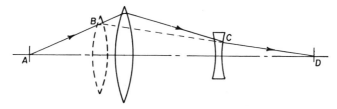

FIG. 28. A two-lens system at finite magnification.

what we know, in a regular (y, u) table. Since the magnification is to be −3, the entering part of the ray must have minus three times the slope of the emerging part, and the two lenses must joint up the two ray sections shown in Fig. 28:

	ϕ		ϕ_a		ϕ_b	
	$-d$			-2		
$l = -2.0$	y		6		2	
	u	-3		(u_b)	1	$l' = 2.0$

Obviously, the intermediate ray slope $u_b = (2 - 6)/(-2) = 2.0$. Then $\phi_a = (u_b + 3)/6 = 5/6 = 0.8333$, and $\phi_b = (1 - u_b)/2 = -0.5$. The required focal lengths are therefore 1.2 and −2 in., respectively.

A glance at Fig. 28 will reveal that any lens system that joins the two sections of the ray will solve the problem; indeed, it could be done with a single lens located at the intersection of AB and CD, shown dashed. For this lens $a = 1.5$, $b = 4.5$, and $f' = 1.125$ in.

G. INSERTION OF THICKNESSES

Having laid out a system of thin lenses to perform some job, we next have to insert suitable thicknesses. A scale drawing of the lenses (assumed equiconvex or equiconcave) will indicate suitable thicknesses, but we must then scale the lenses to their original focal lengths. We next calculate the positions of the principal points of each element, and adjust the air spaces so that the principal-point separations are equal to the original thin-lens separations. If this operation is correctly performed, tracing a paraxial ray from infinity will yield exactly the same focal length and magnification as in the original thin system.

VI. THIN-LENS LAYOUT OF ZOOM SYSTEMS

A zoom lens is one in which the focal length can be varied continuously by moving one or more of the lens components along the axis, the image position being maintained in a fixed plane by some means, either optical or mechanical. If the focal length is varied but the image is not maintained in a fixed plane, the system is said to be varifocal. The latter type is convenient for projection lenses and the lenses on a reflex camera, in which the image focus is observed by the operator before the exposure is made. A true zoom lens must be used in a movie camera or in any situation in which it is necessary to be sure that the focus is maintained during a zoom.

A. MECHANICALLY COMPENSATED ZOOM LENSES

A zoom camera lens is usually composed of a Donders-type afocal system mounted in front of an ordinary camera lens (Fig. 29). To vary the

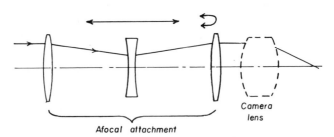

FIG. 29. A mechanically compensated zoom system.

focal length, the middle negative component is moved along the axis, the focal position being maintained by simultaneously moving either the front or the rear component by an in-and-out cam.

Example. Suppose we wish to design a symmetrical Donders telescope in which the magnifying power can be varied over a range of 3 : 1. The magnification of the negative component must therefore vary from $\sqrt{3}$ to $1/\sqrt{3}$, or from 1.732 to 0.577. The focal length of the negative component is found by

$$\text{focal length} = \frac{\text{shift of lens}}{\text{change in magnification}}$$

Suppose $f_a = f_c = 4$ in., and $f_b = -1.0$ in. Then a series of lens locations will be as follows:

Data of middle component			Thin-lens separations		
Magn.	Object dist.	Image dist.	Front	Rear	Image shift
1.732	1.577	−2.732	2.423	1.268	−0.309
1.4	1.714	−2.400	2.286	1.600	−0.114
1.0	2.000	−2.000	2.000	2.000	0
0.7	2.429	−1.700	1.571	2.300	−0.129
0.577	2.732	−1.577	1.268	2.423	−0.309

The last column, image shift, indicates the required movement of either the front or the rear component of the afocal Donders telescope to maintain the image at infinity, so that the telescope can then be mounted in front of a camera set to receive parallel light. Focusing on a near object must be performed by moving the front component axially; otherwise the zoom law will not hold for a close object. Of course, if this is to be a projection lens, there is no need to maintain the afocal condition nor to provide any focusing adjustment for near objects.

The focal length of the camera lens attached to the rear of the Donders telescope can have any value, and it is generally best to use as large an afocal attachment as possible to reduce the aberrations. The early zoom lenses of this type were equipped with simple achromatic doublets for the zoom components.

B. A THREE-LENS ZOOM

In this system once more we have three components, plus–minus–plus, with no fixed lens in the rear (Fig. 30). The first lens is fixed, and the second

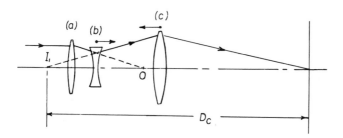

FIG. 30. Layout of a three-lens zoom with mechanical compensation.

and third lenses move in opposite directions. The focal length of the system is equal to the focal length of lens a multiplied by the magnifications of lenses b and c. It is therefore highly desirable that b and c should both magnify or both demagnify together; otherwise the action of c will tend to undo the action of b.

When the negative lens b is at unit magnification, the image I_1 will lie as far to the right as possible. When b moves to the right its magnification will increase, and while this is occurring lens c should be moved to the left so that its magnification will also increase. The computation procedure is simple. Lens a being fixed, its image is also fixed at O. For each position of lens b, its image can be located, and so the object-to-image distance D_c for lens c can be found. Equation (25c) is then employed to calculate m_c and hence the conjugate distances of the third lens.

Example. Let $f_a = 3.0$ with a very distant object, $f_b = -1.0$, and $f_c = 2.7$. The distance from lens a to the image plane is to be 10.0. Four typical positions of the lenses are indicated as follows:

m_b	$1/m_b$	l_b	Separation ab	l'_b	D_c	m_c	l_c	Separation bc	l'_c	Focal length
1.0	1.00	2.00	1.00	−2.0	11.00	1.3117	4.7584	2.7584	6.2416	3.935
1.5	.67	1.67	1.33	−2.5	11.17	1.4426	4.5716	2.0716	6.5951	6.492
2.0	.50	1.50	1.50	−3.0	11.50	1.6550	4.3314	1.3314	7.1686	9.930
2.3	.435	1.435	1.565	−3.3	11.735	1.7864	4.2114	.9114	7.5234	12.326

The focal length range is thus just over $3:1$, although the range of the negative-lens magnification is only $2.3:1$. The motions of the two lenses are indicated in Fig. 31. The focal length of lens a can be anything, and the

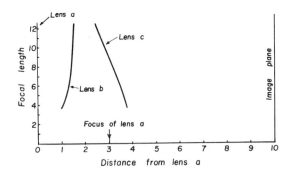

FIG. 31. Lens motions in a three-lens zoom system.

original object distance can be anything, but the image produced by lens a must lie at 7 units in front of the final image plane for these data to be applicable. This type of zoom system is used in a zoom microscope, the objective lens alone producing a virtual object at the final image position.

C. A THREE-LENS OPTICALLY COMPENSATED ZOOM SYSTEM

This system was introduced in 1949 by Cuvillier[2] under the name Pan-Cinor. Two moving lenses are coupled together with a fixed lens between them. Generally the coupled lenses are both positive and the fixed lens is negative, but other arrangements are possible. If the powers and separations of the lenses are properly chosen, then the image will remain virtually fixed while the outer lenses are moved, without any need for a cam, whence the name " optical compensation." To focus on a close object, it is necessary to move the inside negative lens or to vary the separation of the two moving lenses.

The thin-lens predesign of such a system is straightforward, although the algebra involved is complicated. In Fig. 32 we see the system in its initial

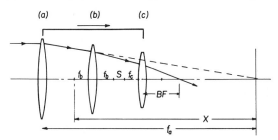

FIG. 32. Layout of a three-lens optically compensated zoom system.

configuration. The separation of adjacent focal points of lenses a and b is X, as shown, and the separation of adjacent foci of lenses b and c is S. Then we can draw up a table of the three lenses, and trace a paraxial ray by the $(y - u)$ method:

ϕ $-d$		$1/f_a$ 	$-(f_a+f_b-X)$	$1/f_b$ 	$-(f_b+f_c+S)$	$1/f_c$ 	
y		1		$(X-f_b)/f_a$		$-\dfrac{f_b^2 + XS + Xf_c}{f_a f_b}$	
u	0		$1/f_a$		$X/f_a f_b$	$-\dfrac{f_b^2 + XS}{f_a f_b f_c}$	

[2] R. H. R. Cuvillier, Le Pan-Cinor et ses applications, *La Tech. Cinemat.* **21**, 73 (1950); also U.S. Patent 2,566,485, filed January, 1950.

The initial focal length is therefore

$$-f_a f_b f_c / (f_b^2 + XS) \tag{27}$$

and the initial back focus is $f_c + f_c^2 X / (f_b^2 + XS)$. Note that the initial back focus is independent of f_a.

Suppose we now move the zoom section (lens a plus lens c) to the right by a distance D. Then X and S will be both increased by D, but to hold the image in a fixed position we require the back focus to be reduced by D. Thus

$$D = \text{(initial back focus)} - \text{(new back focus)}$$

$$= \left[f_c + \frac{f_c^2 X}{f_b^2 + XS} \right] - \left[f_c + \frac{f_c^2 (X + D)}{f_b^2 + (X + D)(S + D)} \right] \tag{28}$$

from which we get

$$f_b^4 + (f_b^2 + SX)(S + D)(X + D) + f_b^2(f_c^2 + SX) - f_c^2 X(X + D) = 0 \tag{29a}$$

Now, for this system to be an effective zoom, we require the image plane to lie in a fixed position for a shift D and also for a shift $2D$. Substituting $2D$ for D in Eq. (29a) gives

$$f_b^4 + (f_b^2 + SX)(S + 2D)(X + 2D) + f_b^2(f_c^2 + SX) - f_c^2 X(X + 2D) = 0 \tag{29b}$$

Subtracting (29a) from (29b) gives

$$f_c^2 = \frac{(f_b^2 + SX)(X + S + 3D)}{X}$$

and substituting this into (28) gives

$$f_b^2 = \frac{X(X + D)(X + 2D)}{S + 2X + 3D} \tag{30}$$

Thus for any set of values for X, S, and D we can solve for the powers of the two lenses b and c. However, we can simplify these expressions by introducing the "zoom range" R, which is the ratio of the initial to the final focal length. Using Eq. (27) we see that

$$R = \frac{f_b^2 + (X + 2D)(S + 2D)}{f_b^2 + XS}$$

which gives us

$$f_b^2 = \frac{(X + 2D)(S + 2D) - RXS}{R - 1} \tag{31}$$

Combining Eqs. (30) and (31) we eliminate f_b and solve for S as a function of R, X, and D:

$$S^2[X(1 - R) + 2D] + S[2X^2(1 - R) + 3DX(3 - R) + 10D^2]$$
$$- (X + 2D)[X(R - 1)(X + D) - 2D(2X + 3D)] = 0$$

For simplicity we can now normalize the system by writing $D = 1$, and then solving for S,

$$S = \frac{2X^2(R - 1) + 3X(R - 3) - 10 \pm [X(R + 1) + 2]}{2X(1 - R) + 4}$$

It will be found that the negative sign of the root gives useful systems, for which

$$S = \frac{X^2(R - 1) + X(R - 5) - 6}{2 - X(R - 1)} \tag{32}$$

Then

$$f_b^2 = \frac{X + 1}{R - 1}(XR - X - 2), \qquad f_c^2 = \frac{4R}{R - 1} \cdot \frac{2 + X + XR}{(2 + X - XR)^2} \tag{33}$$

If R is greater than 1, the moving lenses will be positive, and if R is less than 1, the moving lenses will be negative. In order that the rear air space shall be positive, where $d_b' = (f_b + f_c + S)$, we must select reasonable starting values for X. Approximate suitable values are

R:	5	4	3	2		0.5	0.4	0.3	0.2
X:	1.3	1.7	2.4	4.5		-7.0	-5.5	-4.5	-3.8

Example. Suppose we wish to lay out an optically compensated zoom having $R = 3$, with $X = 2.2$. Then Eqs. (32) and (33) give

$$S = 0.3, \qquad f_b^2 = 3.84, \qquad f_c^2 = 11.25$$

Since R is greater than 1, the two moving lenses will be positive and the fixed lens will be negative. Taking square roots gives

$$f_b = -1.95959 \qquad \text{and} \qquad f_c = 3.35410$$

Assuming that the initial separation between lenses a and b is to be 3.0, we find that the focal length of the front lens must be 7.15959, and the rear air space d_b' will be initially 1.69451. Using the $(y - u)$ method, we calculate the following data:

Shift of zoom components		Back focus	Image shift	Focal length
(Initial	−0.5	8.81839	−0.5357	
position)	0	8.85410	0	10.457
	0.5	8.41660	0.0625	
$D =$	1.0	7.85410	0	5.882
	1.5	7.31839	−0.0357	
	2.0	6.85410	0	3.486
	2.5	6.46440	0.1103	

It is clear that the image plane passes through the three designated positions corresponding to $D = 0$, 1, and 2, but it departs from that plane for all other values of D. These departures, commonly called loops, will be very noticeable if the system is made in a large size, but they can be rendered negligible if the zoom system is made fairly large and is used in front of a small fixed lens of considerable power, as on an 8 mm movie camera. It will be noticed, too, that the law connecting image distance with zoom movement is a cubic (Fig. 33).

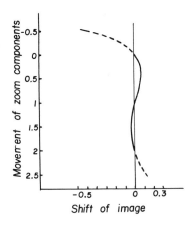

FIG. 33. Image motion with a three-lens optically compensated zoom system.

D. A FOUR-LENS OPTICALLY COMPENSATED ZOOM SYSTEM

We can drastically reduce the sizes of the loops between the in-focus image positions by the use of a four-lens arrangement, as shown in Fig. 34. Here we have a fixed front lens, followed by a pair of moving lenses coupled together with a fixed lens between them. The algebraic solution for the

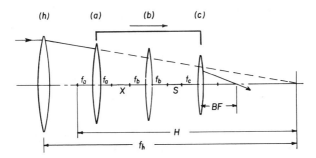

FIG. 34. Layout of a four-lens optically compensated zoom system.

powers and spaces of the four lenses is similar to that already discussed, but it is vastly more complicated.

We now designate the separations between adjacent focal points in the three airspaces by H, X, and S, the X and S serving the same functions as before. The initial lens separations are

$$d'_h = f_h + f_a - H \qquad d'_a = f_a + f_b + X \qquad d'_b = f_b + f_c + S$$

The initial focal length and the initial back focus are

$$\frac{f_h f_a f_b f_c}{f_a^2 S + HXS - f_b^2 H}, \qquad f_c + f_c^2 \left(\frac{f_a^2 + HX}{f_a^2 S + HXS - f_b^2 H} \right)$$

respectively, the denominators being the same in each case. We now shift the moving elements by a distance D, so that S is increased by D, but H, X, and the back focus will be reduced by D. We then substitute $2D$ and $3D$ for D, and after three subtractions we obtain the relationship

$$f_c^2(f_a^2 + HX) + (f_b^2 H - f_a^2 S - HSX)(S - X - H + 6D) = 0 \qquad (34)$$

We can considerably simplify the problem by assuming that the moving lenses a and c have equal power. Then Eq. (34) becomes

$$f_a^4 - f_a^2[-HX + S(S - X - H + 6D)] + H(f_b^2 - SX)(S - X - H + 6D)$$
$$= 0$$

We solve this for f_b^2 in terms of f_a^2, giving

$$f_b^2 = \frac{f_a^2 + HX}{H} \left[S - \frac{f_a^2}{S - X - H + 6D} \right]$$

Substituting f_b^2 in the original equation relating the back focus before and after the zoom shift, and noting that now $S = (X - 3D)$, we get

$$f_a^4 + f_a^2(2H - 3D)(H + X - 3D)$$
$$- H(H - D)(H - 2D)(H - 3D) = 0$$
$$f_b^2 = \frac{f_a^2 + HX}{H}\left[\frac{f_a^2}{H - 3D} + (X - 3D)\right] \tag{35}$$

The focal-length range R, the ratio of the initial to the final focal lengths, is now

$$R = \frac{-f_a^2 X + X(H - 3D)(3D - X) + f_b^2(H - 3D)}{-(HX + f_a^2)(3D - X) - Hf_b^2}$$

This ratio R will be less than 1.0 if the moving lenses are negative.

Example. As an example we shall set up a system having the same range of focal lengths as in the last example, so that we can compare the sizes of the loops. We find that for this case we put $X = 3.5$, $D = 1$, and $H = 10.052343$. The equations just given yield

$$f_a^2 = 25.130858 \quad \text{or} \quad f_a = f_c = -5.0130687$$
$$f_b^2 = 24.380858 \quad \text{or} \quad f_b = 4.937698$$
$$R = 0.333333 \quad (S = 0.5)$$

The initial air spaces are

$$d_h' = f_h + f_a - H = 0.5 \text{ (say)}; \quad \text{hence} \quad f_h = 15.565412$$
$$d_a' = 3.424629$$
$$d_b' = 0.424629$$

We find that using these four lens elements, the overall focal length is negative, and so we must add a fifth lens at the rear to give us the desired positive focal lengths. To compare with the three-lens system in Section VI,C, we set the initial focal length at 3.486, which requires a rear lens having a focal length of 4.490131 located initially 4 units behind the fourth element. Tracing paraxial rays through this system, at a series of zoom positions, gives the tabular data on page 69 (Fig. 35).

It will be noticed that the loops are only about one-fiftieth of their former size, and that the error curve is now a quartic. Obviously, with these very small errors, it would be quite reasonable to design a four-lens zoom of this type covering a much wider range of focal lengths, say 6 : 1 or even more, and this indeed has been done.

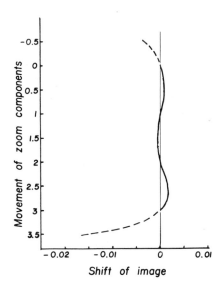

FIG. 35. Image motion with a four-lens optically compensated zoom system.

Shift of zoom components		Back focus	Image shift	Focal length
(Initial	−0.5	6.22805	−0.00383	
position) D	0	6.23188	0	3.486
	0.5	6.23267	0.00079	
$D =$	1.0	6.23188	0	5.026
	1.5	6.23120	−0.00068	
	2.0	6.23188	0	7.253
	2.5	6.23352	0.00164	
	3.0	6.23188	0	10.458
	3.5	6.21538	−0.01650	

E. AN OPTICALLY COMPENSATED ZOOM ENLARGER OR PRINTER

Since the four-lens zoom discussed in Section VI,D can be constructed with two equal positive lenses moving together between three negative lenses, it is obviously possible to remove the two outer negative lenses, leaving a three-lens zoom printer or enlarger system that has a quartic error curve. Equations (35) are now

$$f_a^4 + f_a^2(2H - 3)(H + X - 3) - H(H - 1)(H - 2)(H - 3) = 0$$

$$f_b^2 = \left(X + \frac{f_a^2}{H}\right)\left[\frac{f_a^2}{H-3} + (X - 3)\right] \qquad (36)$$

These can be used to set up a system, taking the positive root for $f_a = f_c$ and the negative root for f_b. The H is the initial distance from the fixed object to the anterior focus of the front lens, the initial object distance being therefore $(H - f_a)$. The initial lens separations are respectively $(f_a + f_b + X)$ and $(f_a + f_b + X - 3)$.

Example. As an example we will design such a zoom system with $H = -8$ and $X = 2$. The above formulas give $f_a = f_c = 6.157183$ and $f_b = -2.667455$. The separations are, respectively, 4.667455 and 1.667455 at the start; they will, of course, be increased or decreased as the zoom elements are moved to change the magnification. The overall distance from object to image is equal to $2(14.157183 + 4.667455) = 37.6493$. Tracing rays by the $(y - u)$ method gives this table of data:

Shift of zoom components		Image distance	Desired image distance	Image shift	Magnification
(Initial	−0.5	17.762025	17.657184	+0.104841	
position)	0	17.157184	17.157184	0	−1.7520
	0.5	16.646875	16.657184	−0.010309	
$D =$	1.0	16.157184	16.157184	0	−1.2071
	1.5	15.661436	15.657184	+0.004252	
	2.0	15.157184	15.157184	0	−0.8285
	2.5	14.652316	14.657184	−0.004868	
	3.0	14.157184	14.157184	0	−0.5708
	3.5	13.680794	13.657184	+0.023610	

The image shift is shown graphically in Fig. 36. It will be noticed that as we are now moving a pair of positive components, the quartic curve is in the opposite direction to that for the previous example, in which we moved a pair of negative lenses.

If it is desired to cover a wider range of magnifications, the value of H should be reduced, and if the lenses come too close together, then X can be somewhat increased. Obviously there is no magic about the size, and if a different object-to-image distance is required, the entire system can be scaled up or down as needed. The fixed negative component is very strong and in practice it is often divided into a close pair of negative achromats, but we leave this up to the designer.

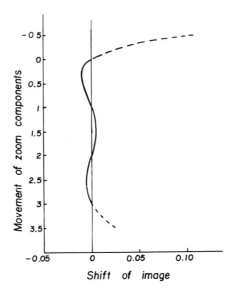

FIG. 36. Image motion for zoom enlarger system.

CHAPTER 4

Chromatic Aberration

Since the refractive index of glass changes with wavelength, it follows that every property of a lens depending on its refractive index will also change with wavelength. This includes the focal length, the back focus, the spherical aberration, field curvature, and all the other aberrations.

I. SPHEROCHROMATISM OF A CEMENTED DOUBLET

For instance, if we take the cemented telescope objective used as an illustration on p. 28 and trace through it a marginal, zonal, and paraxial ray in each of five wavelengths, we obtain the following table of image distances, expressed relative to the paraxial focus in D light:

Wavelength	A' (0.7665)	C (0.6563)	D (0.5893)	F (0.4861)	g (0.4358)
Crown index	1.51179	1.51461	1.517	1.52262	1.52690
Flint index	1.63754	1.64355	1.649	1.66275	1.67408
Marginal $Y = 2$	0.0198	0.0092	0.0080	0.0256	0.0577
Zonal $Y = 1.4$	0.0058	−0.0105	−0.0175	−0.0159	0.0017
Paraxial	0.0320	0.0113	0	−0.0104	−0.0036

These data may be plotted in two ways. First we can plot the longitudinal spherical aberration against aperture, separately in each wavelength (Fig. 37a); and second, we can plot aberration against wavelength for each zone (Fig. 37b). The first set of curves represents the chromatic variation of spherical aberration, or "spherochromatism," and the second set represents the chromatic aberration curves for the three zones. On these curves we notice several specific aberrations.

1. Spherical Aberration (LA')

This is given by $L'_{marginal} - l'_{paraxial}$ in brightest (D) light. It has the value 0.0080 in this example, and is slightly overcorrected.

2. Zonal Aberration (LZA')

This is given by $L'_{zonal} - l'_{paraxial}$ in D light. It has the value -0.0175, and is undercorrected. The best compromise between marginal and zonal aber-

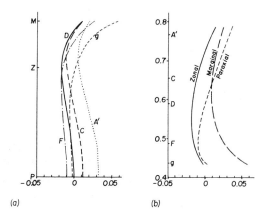

FIG. 37. Spherochromatism ($f' = 12$). (a) Chromatic variation of spherical aberration; (b) chromatic aberration for three zones.

ration for photographic objectives is generally to secure that $LA' + LZA' = 0$, but for visual systems it is better to have $LA' = 0$.

3. Chromatic Aberration (L'_{ch})

This is given by $L'_F - L'_C$, and its magnitude varies from zone to zone (Fig. 38):

Zone	$L'_{ch} = L'_F - L'_C$
Marginal	$+0.0164$
0.7 Zonal	-0.0054
Paraxial	-0.0217

If no zone is specified, we generally refer to the 0.7 zonal chromatic aberration, because zero zonal chromatic aberration is the best compromise for a visual system. Photographic lenses, on the other hand, are generally stopped down somewhat in use, and it is often better to unite the extreme colored foci for about the 0.4 zone instead of the 0.7 zone suggested here.

Chromatic aberration can be represented as a power series of the ray height Y:

$$\text{chromatic aberration} = L'_{ch} = a + bY^2 + cY^4 + \cdots$$

The constant term a is the paraxial or "primary" chromatic aberration. The

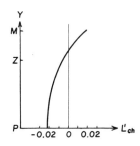

FIG. 38. Variation of chromatic aberration with aperture.

secondary term bY^2 and the tertiary term cY^4 represent the variation of chromatic aberration with aperture as shown in Fig. 38.

4. Secondary Spectrum

This is generally expressed as the distance of the blue or red focus from the combined $C - F$ focus, taken at the height Y at which the C and F curves intersect. In the example on p. 73 the C and F curves intersect at about $Y = 1.5$, and at that height the other wavelengths depart from the combined C and F focus by

Spectrum line:	A'	C	D	F	g
Departure of focus:	0.016	0	−0.007	0	0.016

In the absence of secondary spectrum the curves in Fig. 37b would all be straight lines. The fact that achromatizing a lens for two colors fails to unite the other colors is known as secondary spectrum; it should not be confused with the secondary chromatic aberration mentioned in paragraph 3.

5. Spherochromatism

This is the chromatic variation of spherical aberration and is expressed as the difference between the marginal spherical aberration in F and C light:

$$\text{spherochromatism} = (L' - l')_F - (L' - l')_C$$

$$= (L'_F - L'_C) - (l'_F - l'_C)$$

$$= \text{marginal chromatic aberration}$$

$$- \text{paraxial chromatic aberration}$$

$$= 0.0164 + 0.0217 = 0.0381$$

II. CONTRIBUTION OF A SINGLE SURFACE TO THE PRIMARY CHROMATIC ABERRATION

To determine the contribution of a single spherical surface to the paraxial chromatic aberration of a lens, we write Eq. (18) in F and C light:

$$\frac{n'_F}{l'_F} - \frac{n_F}{l_F} = \frac{n'_F - n_F}{r} \quad \text{and} \quad \frac{n'_C}{l'_C} - \frac{n_C}{l_C} = \frac{n'_C - n_C}{r}$$

Subtracting F from C gives

$$\frac{n'_C}{l'_C} - \frac{n'_F}{l'_F} - \frac{n_C}{l_C} + \frac{n_F}{l_F} = \frac{(n'_C - n'_F) - (n_C - n_F)}{r}$$

We now write $(n_F - n_C) = \Delta n$; hence $n_F = n_C + \Delta n$ and $n'_F = n'_C + \Delta n'$. We insert these values, and we may drop the denominator suffixes without making an excessive approximation error, giving

$$\frac{n'}{l'^2}(l'_F - l'_C) - \frac{n}{l^2}(l_F - l_C) = \Delta n\left(\frac{1}{r} - \frac{1}{l}\right) - \Delta n'\left(\frac{1}{r} - \frac{1}{l'}\right)$$

We next multiply through by y^2, noting that $(1/r - 1/l) = i/y$. Then

$$n'u'^2 L'_{ch} - nu^2 L_{ch} = yi\,\Delta n - yi'\,\Delta n' = yni(\Delta n/n - \Delta n'/n')$$

We write this expression for every surface and add. Much cancellation occurs because of the identities $n'_1 \equiv n_2$, $u'_1 \equiv u_2$, and $L'_{ch_1} \equiv L_{ch_2}$. Hence, if there are k surfaces, we get

$$(n'u'^2 L'_{ch})_k - (nu^2 L_{ch})_1 = \sum yni(\Delta n/n - \Delta n'/n')$$

and dividing through by $(n'u'^2)_k$ gives

$$L'_{ch_k} = L_{ch_1}\left(\frac{n_1 u_1^2}{n'_k u'^2_k}\right) + \sum \frac{yni}{n'_k u'^2_k}\left(\frac{\Delta n}{n} - \frac{\Delta n'}{n'}\right) \tag{37a}$$

The quantity under the summation sign is a surface contribution to the longitudinal paraxial chromatic aberration. Thus we can write

$$L'_{ch}C = \frac{yni}{n'_k u'^2_k}\left(\frac{\Delta n}{n} - \frac{\Delta n'}{n'}\right) \tag{37b}$$

The chromatic aberration of the object, if any, is transferred to the image by the ordinary longitudinal magnification rule and added to the aberration arising at the lens surfaces.

In Table VI we have used these formulas to calculate the paraxial chromatic aberration contributions of the three surfaces of the cemented doublet already used several times. The sum of the contributions is -0.021653. For comparison, we note from p. 73 that $l_F' - l_C' = -0.0217$. The agreement between this contribution formula and actual paraxial ray tracing is extremely close in spite of the various small approximations that we made in deriving the formula.

TABLE VI

PRIMARY CHROMATIC ABERRATION CONTRIBUTIONS

y	2		1.903148	1.880973	
n	1		1.517	1.649	
i	0.270654		-0.459757	-0.171386	
$1/u_k'^2$	36		36	36	
$n_F - n_C = \Delta n$	0	0.00801		0.01920	0
$\Delta n/n$	0	0.005280		0.011643	0
$(\Delta n/n - \Delta n'/n')$	-0.005280		-0.006363	0.011643	
$L_{ch}'C$	-0.102891		0.304054	$-0.222816 \quad \sum = -0.021653$	

III. CONTRIBUTION OF A THIN ELEMENT IN A SYSTEM TO THE PARAXIAL CHROMATIC ABERRATION

The classical relation between object and image distances for a thin lens is

$$\frac{1}{l'} = \frac{1}{l} + (n-1)c$$

We write this in F and C light and subtract C from F. This gives

$$\frac{l_C' - l_F'}{l'^2} - \frac{l_C - l_F}{l^2} = (n_F - n_C)c = \frac{1}{fV} \tag{38}$$

Multiplying by $(-y^2)$ gives

$$L_{ch}'\left(\frac{y^2}{l'^2}\right) - L_{ch}\left(\frac{y^2}{l^2}\right) = -\frac{y^2}{fV} \quad \text{or} \quad L_{ch}'u'^2 - L_{ch}u^2 = -\frac{y^2}{fV}$$

We write this expression for each thin element in the system and add up. After much cancellation, and assuming that there are k elements in the system, we get

$$L_{ch_k}'u_k'^2 - L_{ch_1}u_1^2 = -\sum y^2/fV$$

Finally, dividing through by $u_k'^2$ gives an expression for the chromatic aberration of the image as

$$L_{\mathrm{ch}_k}' = L_{\mathrm{ch}_1}\left(\frac{u_1}{u_k'}\right)^2 - \frac{1}{u_k'^2}\sum\frac{y^2}{fV} \qquad (39)$$

In these expressions, f refers to the focal length of each individual thin element, and V refers to its Abbe number or reciprocal dispersive power,

$$V = \frac{n_D - 1}{n_F - n_C}$$

The magnitude of V varies from 25 for a very dense flint to about 75 for an extra light crown. Every type of optical glass can thus be represented by a point on a chart connecting the mean refractive index n_D with the V number

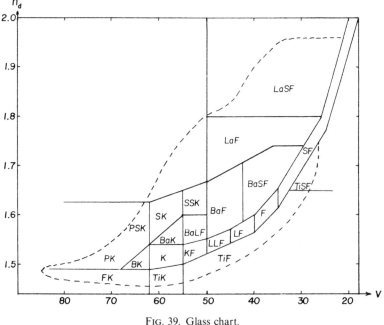

FIG. 39. Glass chart.

(Fig. 39). This diagram shows the approximate ranges of different types of glass, taken from Schott's 1973 catalog. The vertical line at $V = 50$ divides the so-called crown (*kron* in German) and flint types, although these names have long lost any significance. However, we still use the terms loosely to represent glasses having relatively low and high dispersive powers. The narrow band of crowns, light flints, flints, and dense flints contains all the older soda-lime–silica glasses having a progressively increasing lead content. Above this band comes, first, the barium glasses and then (since 1938) a wide range of lanthanum or rare-earth glasses. Within the last few years some titanium flints have been introduced, which fall below the old crown–flint line. At the far left are found some recently introduced fluor and phosphate crowns, some of which have extreme properties. As optical glasses vary enormously in price, from a few dollars up to $300 a pound, the lens designer must watch the price catalog very carefully when selecting glasses to be used in any particular lens.

Returning to Eq. (39) we see that the paraxial chromatic aberration of an isolated single thin lens in air is given by

$$L'_{ch} = -\frac{y^2}{fV}\left(\frac{l'^2}{y^2}\right) = -\frac{l'^2}{fV}$$

and if the object is very distant, this becomes merely

$$L'_{ch} = -f/V$$

The chromatic aberration of a single thin lens with a distant object is therefore equal to the focal length of the lens divided by the V number of the glass. It thus falls between 1/25 and 1/75 of the focal length, depending on the type of glass used in its construction.

For a thin system of lenses in close contact, we can write Eq. (38) for each element and then add. This gives

$$\left[\frac{L'_{ch}}{l'^2} - \frac{L_{ch}}{l^2}\right] = -\sum c\,\Delta n = -\sum\frac{\phi}{V}$$

The quantity on the left-hand side we call the chromatic residual R, which is zero for an achromatic lens with a real object. If the total power of the thin-lens system is Φ, then

$$\Phi = \sum(\phi) = \sum(Vc\,\Delta n) \qquad \text{and} \qquad R = -\sum(c\,\Delta n)$$

For the very common case of a thin doublet, these equations become

$$1/F' = \Phi = V_a(c\,\Delta N)_a + V_b(c\,\Delta N)_b$$

$$-R = (c\,\Delta N)_a + (c\,\Delta N)_b$$

Solving for c_a and c_b gives the important relationships

$$c_a = \frac{1}{F'(V_a - V_b)\,\Delta N_a} + \frac{RV_b}{(V_a - V_b)\,\Delta N_a}$$

$$c_b = \frac{1}{F'(V_b - V_a)\,\Delta N_b} + \frac{RV_a}{(V_b - V_a)\,\Delta N_b}$$

(40)

These are the so-called (c_a, c_b) equations which are used to start the design of any thin achromatic doublet. In most practical cases the chromatic residual R is zero and then only the first terms need be considered. The condition for achromatism is then independent of the object distance, and we say that achromatism of a thin system is "stable" with regard to object distance.

Since for a thin lens $f' = 1/c(N - 1)$, we can convert the c_a, c_b formulas into the corresponding focal-length formulas for $R = 0$:

$$f'_a = F'\left(\frac{V_a - V_b}{V_a}\right) \quad \text{and} \quad f'_b = F'\left(\frac{V_b - V_a}{V_b}\right) \tag{41}$$

For an ordinary crown glass with $V_a = 60$ and an ordinary flint with $V_b = 36$, we have $V_a - V_b = 24$, and the power of the crown element is seen to be 2.5 times the power of the combination, while the power of the flint is -1.5 times as strong as the doublet. Hence, to achromatize a thin lens requires the use of a crown element 2.5 times as strong as the element itself (Fig. 40).

FIG. 40. An $f/3.5$ single lens and an achromat of the same focal length.

Consequently, although a single lens of aperture $f/1$ is not excessively strong, it is virtually impossible to make an achromat of aperture much over $f/1.5$.

It is important to note that chromatic aberration depends only on lens powers and not at all on bending or surface configuration. Attempts to modify the chromatic correction of a lens by hand rubbing on one of the surfaces are generally quite unsuccessful, because it requires a very large change in the lens to produce a noticeable change in the chromatic aberration.

IV. PARAXIAL SECONDARY SPECTRUM

We have so far regarded an achromatic lens as one in which the C and F foci are coincident. However, as we have seen, in such a case the D (yellow) focus falls short and the g (blue) focus of the same zone falls long. To determine the magnitude of the paraxial secondary spectrum of a lens in which the paraxial C and F foci coincide, we write the chromatic aberration contribution of a single thin element, for two wavelengths λ and F, as

$$L'_{ch}C \text{ (for } \lambda \text{ to } F) = -\frac{y^2 c}{u'^2_k}(n_\lambda - n_F) = L'_{ch}C\left(\frac{n_\lambda - n_F}{n_F - n_C}\right)$$

The quantity in parentheses is another intrinsic property of the glass, known as the partial dispersion ratio from λ to F. It is generally written $P_{\lambda F}$. Hence for any succession of thin elements

$$l'_\lambda - l'_F = \sum P_{\lambda F}(L'_{\text{ch}} C) = -\frac{1}{u'^2_k}\sum \frac{Py^2}{f'V} \tag{42}$$

For the case of a thin achromatic doublet, y is the same for both elements, and Eq. (41) shows that $f'_a V_a = -f'_b V_b = F'(V_a - V_b)$; hence

$$l'_\lambda - l'_F = -F'\left(\frac{P_a - P_b}{V_a - V_b}\right) \tag{43}$$

For any particular pair of wavelengths, say F and g, we can plot the available types of glass on a graph connecting P_{gF} with V, as in Fig. 41. All the

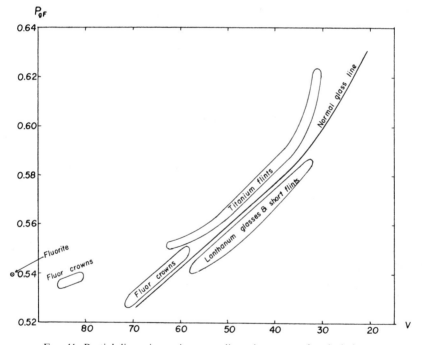

FIG. 41. Partial dispersion ratio versus dispersive power of optical glasses.

common types of glass lie on a straight line that rises slightly for the very dense flints. Below this line come the "short" glasses, which exhibit an unusually short blue end to the spectrum; these are mostly lanthanum crowns and so-called short flints (KzF and KzFS types). Above the line are a few "long" crowns with an unusually stretched blue spectrum: this region also contains some plastics and crystals such as fluorite. The titanium flints also fall above the line, as can be seen.

If we join the points belonging to the two glasses used to make an achromatic doublet, the slope of the line is given by

$$\tan \psi = \frac{P_a - P_b}{V_a - V_b}$$

and clearly the secondary spectrum is given by $F' \tan \psi$. The fact that most of the ordinary glasses lie on a straight line indicates that the secondary spectrum will be about the same for any reasonable selection of glass types. For example, if we choose Schott's K-5 and F-4, we find that the secondary spectrum for a number of wavelengths, assuming a focal length of 10, is

		$A' - F$	$D - F$	$g - F$	$h - F$
K-5	$V_a = 59.63$	$P_a = -1.3470$	-0.7043	0.5457	0.9989
F-4	$V_b = 36.61$	$P_b = -1.3207$	-0.7150	0.5825	1.0891
		$l'_\lambda - l'_F = 0.01142$	-0.00465	0.01599	0.03918

We can reduce the secondary spectrum by choosing a long crown, such as fluorite,[1] with a matching barium crown glass as the flint element:

		$A' - F$	$D - F$	$g - F$	$h - F$
Fluorite	$V_a = 94.93$	$P_a = -1.3457$	-0.7046	0.5383	0.9803
SK-20	$V_b = 61.22$	$P_b = -1.3501$	-0.7024	0.5383	0.9836
		$l'_\lambda - l'_F = -0.0013$	$+0.0007$	0	$+0.0010$

This amount of secondary spectrum is obviously vastly smaller than we found using ordinary glasses. On the other hand, we shall increase the secondary spectrum if we use a normal crown with a long flint such as a titanium glass:

		$D - F$	$g - F$
K-5	$V_a = 59.63$	$P_a = -0.7043$	0.5457
TiF-6	$V_b = 30.97$	$P_b = -0.7220$	0.6220
		$l'_\lambda - l'_F = -0.0062$	$+0.0266$

---

[1] I. H. Malitson, A redetermination of some optical properties of calcium fluoride, *Appl. Opt.* **2**, 1103 (1963).

These residuals are about 1.5 times as large as for the normal glasses listed above.

In view of the apparent inevitability of secondary spectrum, we may wonder why it is necessary to achromatize a lens at all. This question will be immediately answered by a glance at Fig. 42, where we have plotted to the

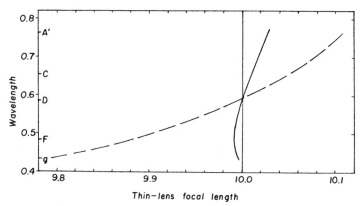

FIG. 42. Comparison of an achromat with a single lens.

same scale the paraxial secondary spectrum curve of the example in Fig. 37b and the corresponding curve for a simple lens of crown glass, K-5.

If a lens has a small residual of primary chromatic aberration, the secondary spectrum curve will become tilted. The three curves sketched in Fig. 43 show what happens in this case. It will be noticed that when the chromatic aberration is undercorrected the wavelength of the minimum focus moves toward the blue; for a $C - F$ achromat it falls in the yellow-green; and for an overcorrected lens it rises up toward the red. Lenses for use in the near infrared must be decidedly overcorrected, whereas lenses intended for use with color-blind film or bromide paper should be chromatically undercorrected.

In any achromat of high aperture the spherochromatism and other residual aberrations are likely to be so much greater than the secondary spectrum that the latter can often be completely ignored. However, in a low-aperture lens of long focal length such as an astronomical telescope objective, in which the other aberration residuals are either corrected in the design or removed by hand figuring, the secondary spectrum may well be the only outstanding residual, and it is then important to consider the possibility of removing it by a suitable choice of special types of glass. Fluorite is commonly used in microscope objectives for this purpose.

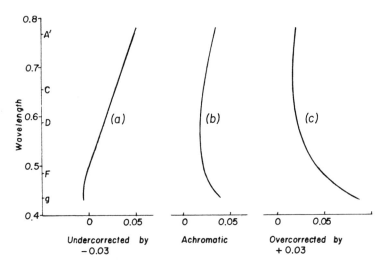

FIG. 43. Effect of chromatic residual in a cemented doublet ($f' = 10$).

V. PREDESIGN OF A THIN THREE-LENS APOCHROMAT

As there are many practical objections to the use of fluorite as a means for reducing secondary spectrum, it is often preferred to unite three wavelengths at a common focus by the use of three different types of glass.

For a thin system with a very distant object, which is achromatized and also corrected for secondary spectrum, we have the three relationships

$$\sum (Vc\,\Delta n) = \Phi \qquad \text{(power)}$$

$$\sum (c\,\Delta n) = 0 \qquad \text{(achromatism)}$$

$$\sum (Pc\,\Delta n) = 0 \qquad \text{(secondary spectrum)}$$

For a thin three-lens apochromat, these equations can be extended to

$$V_a(c_a \cdot \Delta n_a) + V_b(c_b\,\Delta n_b) + V_c(c_c\,\Delta n_c) = \Phi$$

$$(c_a\,\Delta n_a) + (c_b\,\Delta n_b) + (c_c\,\Delta n_c) = 0$$

$$P_a(c_a\,\Delta n_a) + P_b(c_b\,\Delta n_b) + P_c(c_c\,\Delta n_c) = 0$$

These can be solved for the three curvatures as follows:

$$c_a = \frac{1}{F'E(V_a - V_c)}\left(\frac{P_b - P_c}{\Delta n_a}\right)$$

$$c_b = \frac{1}{F'E(V_a - V_c)}\left(\frac{P_c - P_a}{\Delta n_b}\right)$$

$$c_c = \frac{1}{F'E(V_a - V_c)}\left(\frac{P_a - P_b}{\Delta n_c}\right)$$

Note the cyclic order of the terms, and that the coefficient in front of the parentheses is the same in each case. The meaning of E is as follows: If we plot the three chosen glasses on the $P - V$ graph shown in Fig. 41 and then join the three points to form a triangle, E is the vertical distance of the middle glass from the line joining the two outer glasses, E being considered negative if the middle glass falls below the line. Algebraically E is computed by

$$E = \frac{V_a(P_b - P_c) + V_b(P_c - P_a) + V_c(P_a - P_b)}{V_a - V_c}$$

Since E appears in the denominator of all three c expressions, it is clear that the lenses will become infinitely strong if all three glasses fall on a straight line, and conversely, all the elements will become as weak as possible if we select glass types having a large E value. The most usual choice is some kind of crown for lens a, a very dense flint for lens c, and a short flint or lanthanum crown for the intermediate lens b. Once the three curvatures have been calculated, the actual lenses can be assembled in any order.

As an example in the use of these formulas, we will select three glasses forming a wide triangle on the graph of Fig. 44, namely, Schott's FK-6,

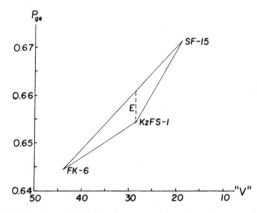

FIG. 44. $P - V$ graph of the glasses used in a three-lens apochromat.

KzFS-1, and SF-15. Since the calculated curvatures depend on the differences between the P numbers, it is necessary to know these to many decimal places, requiring a knowledge of the individual refractive indices to about seven decimals, beyond the capability of any measurement procedure. We therefore use the six-term interpolation formulas given in the current Schott catalog to calculate refractive indices to the required precision. Failure to do this results in such scattered points that it is impossible to plot a smooth color curve for the completed design.

To unite the C, e, and g lines at a common focus, we use the following:

Lens	Glass	n_e	$\Delta n = n_g - n_C$	$n_g - n_e$	P_{ge}	"V"
a	FK-6	1.4478604	0.0101615	0.0065470	0.6442946	44.07424
b	KzFS-1	1.6163841	0.0215499	0.0140995	0.6542721	28.60264
c	SF-15	1.7044410	0.0371905	0.0249596	0.6711283	18.94142

Using these somewhat artificial numbers, we calculate the value of E as -0.0065411, giving $c_a = 1.0090432$, $c_b = -0.7574313$, and $c_c = 0.1631915$. Using other refractive indices also calculated by the interpolation formulas, we can plot the color curve, which duly passes through the points for C, e, and g as required (Fig. 45). It will be seen that there is a very small residual

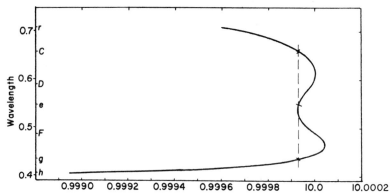

FIG. 45. Tertiary spectrum of a thin three-lens apochromat (the C, e, and g lines are brought to a common focus).

of tertiary spectrum, the foci for the D and F lines being slightly back, while the two ends of the curve move rapidly inwards toward the lens.

This system should be called a "superachromat," since the three glasses satisfy Herzberger's condition[2] for the union of four wavelengths at a

---

[2] M. Herzberger, Colour correction in optical systems and a new dispersion formula, Opt. Acta (London) 6, 197 (1959).

common focus. Failure to meet this condition generally ends up with three united wavelengths in the visible spectrum with the fourth wavelength falling far out into the infrared.

By way of comparison we see that the residual tertiary spectrum in the lens of Fig. 45 is only about 1/100 the secondary spectrum of an ordinary doublet such as that in Fig. 37b. In Chapter 6, Section IV, the design of the apochromat will be completed, after inserting suitable thicknesses and choosing such a shape that the spherical aberration is also corrected.

VI. THE SEPARATED THIN-LENS ACHROMAT (DIALYTE)

If the two elements in an achromat are separated by a small finite distance, the lens is known as a *dialyte*, and we find that the flint element particularly must be considerably strengthened. In Fig. 46 we show the two

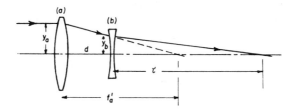

FIG. 46. The dialyte lens.

components of a thin achromatic doublet separated by a finite distance d. It is convenient to express d as a fraction k of the focal length of the crown lens, i.e., $k = d/f'_a$. Since the chromatic aberration contributions of the two elements in an achromat must add up to zero, we see that

$$\frac{y_a^2}{f_a V_a} + \frac{y_b^2}{f_b V_b} = 0$$

but by Fig. 46 it is clear that $y_b = y_a(f'_a - d)/f'_a$, or $y_b = y_a(1 - k)$. Combining these gives

$$f_b V_b = -f_a V_a(1 - k)^2$$

Since the system must have a specified focal length F', we have

$$\frac{1}{F'} = \frac{1}{f'_a} + \frac{1}{f'_b} - \frac{d}{f'_a f'_b} = \frac{1}{f'_a} + \frac{1 - k}{f'_b}$$

Combining the last two relationships gives the focal lengths of the two components as

$$f'_a = F' \left[1 - \frac{V_b}{V_a(1-k)} \right], \quad f'_b = F'(1-k) \left[1 - \frac{V_a(1-k)}{V_b} \right] \quad (44)$$

As an example, let us assume that $V_a = 60$ and $V_b = 36$. The two focal lengths are then related to the value of k as follows:

k:	0	0.1	0.2	0.3
f'_a	$0.4F'$	$0.333F'$	$0.25F'$	$0.143F'$
f'_b	$-0.667F'$	$-0.45F'$	$-0.267F'$	$-0.117F'$
d	0	$0.033F'$	$0.05F'$	$0.043F'$

As k is increased the powers of both lenses become greater, but the power of the negative lens increases more rapidly than that of the positive lens. The two powers become identical at $k = 0.225$, at which point both the focal lengths are $0.225F'$. This property of an achromatic dialyte is employed with great effect in the predesign of a dialyte-type four-element photographic objective (see Chapter 12, Section II). The limiting value of k occurs when $V_a(1 - k) = V_b$, when both elements become infinitely strong. In the present example this is when $k = 0.4$.

A. SECONDARY SPECTRUM OF A DIALYTE

In Eq. (42) we saw that for a succession of thin elements

$$l'_\lambda - l'_F = -\frac{1}{u'^2_k} \sum \frac{Py^2}{f'V}$$

Substituting in this the values of f'_a, f'_b, and y_b for a dialyte gives

$$l'_\lambda - l'_F = -\frac{F'(1-k)}{V_a(1-k) - V_b} (P_a - P_b) \quad (45)$$

When $k = 0$ for a cemented lens this, of course, degenerates to Eq. (43).

Actually, neither the achromatism relation (44) nor the secondary spectrum expression (45) is strictly correct, because in their derivation we assumed that $y_b = y_a(1 - k)$ for all wavelengths. Because of the dispersion of the front element and the finite separation between the elements, it turns out that y_b is a little smaller in blue than in red light. Thus, a dialyte made in accordance with Eq. (44) turns out to be slightly overcorrected chromatically, requiring a slight decrease in the power of the flint element to achromatize. For the same reason the secondary spectrum turns out to be slightly less than the amount given by Eq. (45).

To illustrate, suppose we design a thin-lens dialyte using these glasses:

Glass	n_C	n_e	n_F	$\Delta n = n_F - n_C$	$V_e = \dfrac{n_e - 1}{n_F - n_C}$
K-3	1.51554	1.52031	1.52433	0.00879	59.193
F-4	1.61164	1.62058	1.62848	0.01684	36.852
					$V_a - V_b = \overline{22.341}$

Using formulas (44) we find that, for $F' = 10$ and $k = 0.2$;

$$f'_a = 2.21783, \quad \text{whence} \quad c_a = 0.866581 \quad [\text{because } c = 1/f'(n-1)]$$

$$f'_b = -2.27991, \quad \text{whence} \quad c_b = -0.706781$$

$$d = 0.443566$$

Tracing paraxial rays in C, e, and F through this system using the ordinary thin-lens $(y - u)$ method gives

$$l'_C = 8.008133, \quad l'_e = 8.0, \quad l'_F = 8.008431$$

There is thus a small residual of paraxial chromatic aberration of magnitude 0.000298 in the overcorrected sense. To remove this, we must weaken the flint element slightly, to $c_b = -0.706449$, which gives

$$l'_C = 7.994955, \quad l'_e = 7.986857, \quad l'_F = 7.994962$$

The $F - C$ aberration is now corrected, and the e image lies closer to the lens by an amount of secondary spectrum equal to -0.008103. A thin cemented achromat of the same focal length made of the same glasses has a $D - F$ secondary spectrum of -0.004820, only about half that of the dialyte (see Fig. 47).

B. A One-Glass Achromat

It has been known for a long time that it is actually possible to design an air-spaced achromat using only one kind of glass. If we write $V_a = V_b$ in Eq. (44) we obtain the focal lengths of the two elements of a one-glass achromat as follows:

$$f'_a = \frac{kF'}{k-1}, \quad f'_b = -kF'(k-1), \quad d = kf'_a, \quad l' = -F'(k-1)$$

$$(46)$$

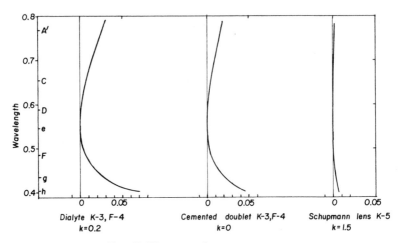

Fig. 47. Three secondary spectrum curves.

once again assuming a very distant object and thin lenses.

Since the air-space d must always be positive, we see that k must have the same sign as f'_a, and $k - 1$ must have the same sign as F'.

For a positive lens, k must be greater than 1.0, which makes for a very long system (see Fig. 48). This is known as a Schupmann lens, but it is

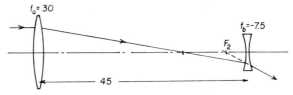

Fig. 48. A Schupmann lens ($f' = 10$).

seldom used because the image is inside.

For a negative system $k - 1$ must be negative, so that k must be less than 1.0. If the front element is positive, k must be positive and thus must lie between 0 and 1. This gives a compact system (see Fig. 49a). If the front element is negative, k must be negative but may have any value. If k is small the system is compact, but if k is large the system becomes very long (Fig. 49b,c). A negative one-glass achromat could, for instance, be used in the rear member of a telephoto lens.

When designing a Schupmann dialyte, the colored rays become separated at the rear component because of the long air space, and so the simple formulas fail to give a perfect achromat and we must readjust the rear lens power for achromatism. Similarly, since both elements have the same dispersion, we might expect the secondary spectrum to be zero, but it is

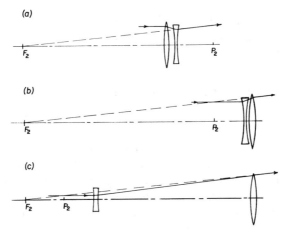

FIG. 49. Negative one-glass achromatic dialytes $(f' = -10)$. (a) $k = 0.2$, $f_a = 2.5$. $f_b = -1.6$, $d = 05$; (b) $k = -0.2$, $f_a = -1.66$, $f_b = 2.4$, $d = 0.33$; (c) $k = -5.0$, $f_a = -8.33$, $f_b = 300$, $d = 41.7$.

actually slightly undercorrected.

As an example, we will design a Schupmann dialyte of focal length 10.0 using K-5 glass in both elements. We take $k = 1.5$, and Eq. (46) tells us that

$$f'_a = 30, \qquad f'_b = -7.5, \qquad d = 45, \qquad l' = -5$$

The refractive indices of K-5 glass are

$$n_C = 1.51981, \qquad n_D = 1.52240, \qquad n_F = 1.52857$$

Therefore $c_a = 0.063808$ and $c_b = -0.255232$. Tracing paraxial rays in these wavelengths through the thin-lens solution gives

$$l'_F = -4.998653 \qquad \text{and} \qquad l'_C = -4.999746$$

leaving a residual of chromatic aberration of $+0.001093$. To remove this we weaken the flint component to $c_b = -0.250217$, which gives the following back foci for a number of wavelengths:

Wavelength		Back focus	Departure from D
A'	0.7682	-5.06442	$+0.00194$
C	0.6563	-5.06577	$+0.00059$
D	0.5893	-5.06636	0
e	0.5461	-5.06647	-0.00011
F	0.4861	-5.06578	$+0.00058$
g	0.4359	-5.06353	$+0.00283$
h	0.4047	-5.06058	$+0.00578$

These data are plotted in Fig. 47 in comparison with the corresponding secondary spectrum curves for a cemented achromat and a dialyte made with ordinary glasses.

VII. CHROMATIC ABERRATION TOLERANCES

A. A SINGLE LENS

In the seventeenth century astronomers used simple lenses of very long focal length as telescope objectives. In this way they managed to make the chromatic aberration insignificant. The logic behind this procedure is that the chromatic aberration of a simple lens is equal to f/V, while the focal range based on diffraction theory is equal to $\lambda/\sin^2 U' = 4\lambda f^2/D^2$, where D is the diameter of the lens. Assuming that, because of the drop in sensitivity of the eye at the deep red and blue, we may let the chromatic aberration reach twice the focal range, we have

$$f/V = 8\lambda f^2/D^2$$

Working in inches, we find that $\lambda = 1/50,000$ approximately, and if $V = 60$, then our formula tells us that the shortest possible focal length to meet this relation is roughly equal to 100 times the square of the lens diameter in inches. Thus an objective of 4 in. aperture will have an insignificant amount of chromatic aberration if its focal length is greater than 1600 in., or 130 feet.

B. AN ACHROMAT

By s similar logic, we can determine the minimum focal length of an achromatic telescope objective for the secondary spectrum to be invisible to the observer. Now we equate the secondary spectrum in D light to the whole focal range, or

$$f/2200 = 4\lambda f^2/D^2$$

whence $f = 5D^2$ approximately. Consequently a 4-in. achromatic objective will have an insignificant amount of secondary spectrum if its focal length is greater than about 80 in. or $6\frac{1}{2}$ feet. The enormous gain resulting from the process of achromatizing is clearly evident.

VIII. CHROMATIC ABERRATION AT FINITE APERTURE

It is clear from the graphs in Fig. 37 that the chromatic aberration of a lens, expressed as $L'_F - L'_C$, varies across the aperture, and a graph of chromatic aberration against incidence height Y appears in Fig. 38. Thus a normal achromat has some degree of chromatic undercorrection for the paraxial rays and a corresponding degree of chromatic overcorrection for the marginal rays, it being well corrected for the 0.7 zonal rays. To achromatize a finite-aperture lens therefore requires the tracing of zonal rays in the two wavelengths that are to be united at a common focus, and experimentally varying one of the radii until these two foci become coincident.

A. Conrady's $D - d$ Method of Achromatization

In 1904 Conrady[3] suggested a very useful and simple procedure for achromatizing a lens, depending on the fact that in an achromat

$$\sum (D - d) \, \Delta n = 0$$

where D is the distance measured along the traced marginal ray in brightest light from one surface to the next, and d is the axial separation of those surfaces. Δn is the index difference between the two wavelengths that are to be united at a common focus, for the material occupying the space between the two lens surfaces under consideration. Since Δn for air is zero, we need consider only glass lenses in making this calculation. The argument used in deriving this relation is as follows:

Suppose we have a series of rays in one wavelength originating at an axial object point and passing through a lens. Each point in the wavefront will travel along the ray and will eventually emerge from the rear of the lens, the moving wavefront being always orthogonal to the rays (Malus' theorem).

Since the emerging wavefront has the property that light takes the same time to go from the source to every point on the wavefront, we see that

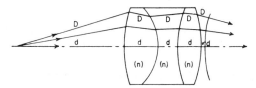

FIG. 50. The emerging wavefront from a lens.

(Fig. 50) time $= \sum (D/v)$, where v is the velocity of light in each section of the ray path of length D. Hence time $= \sum (D/c)(c/v)$ where c is the velocity of light in air. Thus time $= (1/c) \sum (Dn)$ since the refractive index n is equal to the ratio of the velocity of light in air to its velocity in the glass. The $\sum (Dn)$ is

[3] A. E. Conrady, p. 641. See also A. E. Conrady, On the chromatic correction of object glasses (first paper), M. N. Roy. Astron. Soc. 64, 182 (1904).

the length of the optical path along the traced ray, from the original object point to the emerging wavefront, and all points on a given wavefront have the same value of $\sum (Dn)$.

Conrady then proceeded to assume that in a lens having some residual of spherical aberration and spherochromatism, as most lenses do, the best possible state of achromatism occurs when the emerging wavefronts in C and F light (red and blue) cross each other on the axis and at the margin of the lens aperture, as indicated in Fig. 51. Since the C and F wavefronts will

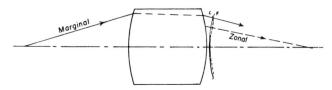

FIG. 51. The emerging wavefronts from an achromat.

then be parallel to each other at about the 0.7 zone, the C and F rays through that zone will lie together and cross the axis at the same point. Under these circumstances

$$\sum (Dn)_C = \sum (Dn)_F$$

along the marginal ray. However, since all points on a wavefront have the same value of $\sum (Dn)$, it is clear that in an achromat

$$\sum (D - d)n_C = \sum (D - d)n_F \qquad \text{or} \qquad \sum (D - d)(n_F - n_C) = 0 \quad (47)$$

This is Conrady's condition for the best possible state of achromatism in a lens that suffers from other residuals of aberration. The presence of spherochromatism, for example, causes the two emerging wavefronts in C and F light to separate between the axis and margin of the aperture, while the presence of spherical aberration causes the wavefronts to assume a noncircular shape.

In stating this condition, we are tacitly assuming that the values of D within all the lens elements are equal for C and F light. This is certainly not true, but we shall make only a very small error if we trace the marginal ray in brightest light, which is usually D or e for $C - F$ achromatism, and calculate the distances D along that ray. The argument breaks down if there is a long air space between unachromatized or only partially achromatized separated components, but in most cases it is surprisingly accurate.

The $D - d$ relation would be impossibly difficult to use if we had to calculate every D value from the original object right up to the emerging wavefront, but the method is saved by the fact that the dispersion $\Delta n = n_F - n_C$ of air is zero. For this reason we must calculate $D - d$ only for those sections of the marginal ray that lie in glass. The length D is found by the usual relation

$$D = (d + X_2 - X_1)/\cos U'_1$$

where $X = r[1 - \cos(U + I)]$ or $X = G \sin(U + I)$, as explained on p. 27. The choice of dispersion values depends on the region of the spectrum in which achromatism is desired. For ordinary visual achromatism, we trace the ray in D or e light and use $\Delta n = n_F - n_C$; for photographic achromatism, we may prefer to trace the marginal ray in F light and use $\Delta n = n_g - n_D$ for the dispersion. The process for interpolating dispersions suggested on p. 14 is of value here.

To achromatize a lens then, we must make the sum $\sum (D - d) \Delta n$ equal to zero by some means or other. Commonly we calculate that value of the last radius of the lens that will accomplish this. Alternatively we may design the lens using any suitable refractive indices, and then at the end search the glass catalog for glass types with dispersion values that will make the $(D - d) \Delta n$ sum zero. To use the first method, suppose that the value of the $D - d$ sum for all the lens elements prior to the last element is Σ_0; then for the last element we must have

$$\sum (D - d) \Delta n = -\Sigma_0$$

We now calculate the X and Y at the next-to-last surface, and knowing the desired value of D in the last element to achieve achromatism, we calculate

$$X_2 = D \cos U'_1 + X_1 - d \quad \text{and} \quad Y_2 = Y_1 - D \sin U'_1$$

(Here the indices 1 and 2 refer to the first and second surfaces of the last element.) The radius of curvature of the last surface is given by

$$r = (X^2 + Y^2)/2X$$

and the problem is solved. As a check on our work, we may wish to trace zonal rays in F and C light through the whole lens; if everything is correct, these rays should cross the axis at the same point in the image space.

B. TOLERANCE FOR $D - d$ SUM

Conrady[4] suggests that in a visual system the tolerance for the $D - d$ sum is about half a wavelength. However, there is no point in achieving perfect achromatism for the 0.7 zonal rays, which the $D - d$ method does, if there is considerable spherochromatism in the lens, since this will swamp the excellent color correction. Therefore we have found that a more reasonable tolerance is about 1% of the contribution of either the crown or the flint

---

[4] A. E. Conrady, p. 647.

element in the lens. If these contributions are small, it indicates that the spherochromatism will be small, and a tight tolerance for the $D - d$ sum is sensible.

Example. As an example in the use of the $D - d$ method, we return to the cemented doublet lens used as a ray-tracing example on p. 28, and compute the $(D - d) \, \Delta n$ sum along the traced marginal ray (Table VII). It will be seen that there is a small residual of the sum, amounting to -0.0000578, which is about 1% of the separate contributions of the crown and flint elements. We must therefore regard this lens as noticeably undercorrected for chromatic aberration. That is the reason why the C and F curves in Fig. 37a cross somewhat above the 0.7 zone of the aperture.

If we wish to achromatize this lens perfectly, we can solve for the last radius by the method described on p. 95. This tells us that a last radius of -16.6527 would make the $D - d$ sum exactly zero. As this radius is decidedly different from the given radius of -16.2225, we see once again that it is necessary to change a lens drastically if we wish to affect the chromatic correction.

TABLE VII

CALCULATION OF THE $(D - d) \, \Delta n$ SUM

c	0.1353271	-0.1931098	-0.0616427	
X	0.2758011	-0.3865582	-0.1149137	
d		1.05	0.4	
$\cos U$		0.9955195	0.9985902	
D		0.3893853	0.6725927	
$D - d$		-0.6606147	0.2725927	
Δn		0.00801	0.01920	
Prod.		-0.0052916	$0.0052338 \quad \sum = -0.0000578$	

As an alternative method for achromatizing, we could calculate what value of Δn for either the crown or the flint glass would be required to eliminate the $D - d$ sum. The numbers in Table VII tell us that we could achromatize with the given crown if we had a flint with $\Delta n = 0.01941$; this represents a V number of 33.43 instead of the given 33.80. Or we could retain the given flint and seek a crown with $\Delta n = 0.00792$; this represents a V number of 65.26 instead of the given 64.54. In both cases the required change in V number is only slightly larger than the normal factory variation in successive glass melts, indicating that the small residual of chromatic aberration in this lens is really almost insignificant.

C. Relation between the $D - d$ Sum and the Ordinary Chromatic Aberration

D. P. Feder[5] has shown that, for any zone of a lens, the vertical displacement in the paraxial focal plane between marginal rays in F and C light is given closely by

$$H'_F - H'_C = \frac{d \sum}{d(\sin U')}$$

where $\sum$ is the sum $\sum (D - d) \Delta n$ calculated along the zonal ray in question, and $\sin U'$ is the emerging slope of the same ray. Thus if we can express $\sum$ as a polynomial of the form

$$\sum = a \sin^2 U' + b \sin^4 U' + c \sin^6 U' \tag{48}$$

then

$$(H'_F - H'_C) = 2a \sin U' + 4b \sin^3 U' + 6c \sin^5 U'$$

By calculating $\sum$ for three zones of a lens, we can solve for the three coefficients a, b, and c, and we then find an excellent agreement with Eq. (49).

A more convenient but only approximate relation between $(H'_F - H'_C)$ and $\sum$ can be found by neglecting the $\sin^6 U'$ term in Eq. (48). When this is done, we can relate the 0.7 zonal chromatic aberration with the marginal $\sum$ in the following way:

Writing

$$S = a \sin^2 U' + b \sin^4 U'$$

for the $D - d$ sum along any zonal ray, we see that if the angle between the C and F rays at any zone is α, then (Fig. 52)

$$L'\alpha = dS/d(\sin U') = 2a \sin U' + 4b \sin^3 U'$$

The longitudinal chromatic aberration for this zone is given approximately by

$$L'_{ch} = L'\alpha/\sin U' = 2a + 4b \sin^2 U'$$

---

[5] D. P. Feder, Conrady's chromatic condition, *J. Res. Nat. Bur. Std.*, **52**, 47 (1954); res. paper 2471.

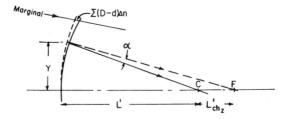

FIG. 52. Relation between the $D - d$ sum and the zonal chromatic aberration.

and hence the 0.7 zonal chromatic aberration will be given by

$$(L'_{ch})_z = 2(a + 2b \sin^2 U'_z)$$

But $\sin U'_z = \sin U'_m/\sqrt{2}$ approximately, and the calculated marginal $\sum (D - d) \Delta n$ sum is

$$\sum = S_m = a \sin^2 U'_m + b \sin^4 U'_m$$

Hence

$$(L'_{ch})_z = 2 \sum/\sin^2 U'_m \qquad (49)$$

As a check on this result, we recall that the residual of $\sum$ in our cemented telescope doublet was -0.0000578 and $\sin U'$ was 0.16659. Therefore, we should expect the zonal chromatic aberration to be -0.00417. By actual ray tracing we find

$$\text{zonal } L'_F = \quad 11.27022$$

$$\text{zonal } L'_C = \quad \underline{11.27523}$$

$$\therefore \ F - C = -0.00501$$

The small discrepancy is due to our having neglected the $\sin^6 U'$ term in the expression for S.

D. PARAXIAL $D - d$ FOR A THIN ELEMENT

We can readily reduce the $D - d$ expression to its paraxial form for a single thin lens element. In this case the length D in the paraxial region becomes

$$D = d + X_2 - X_1 = d + \frac{Y^2}{2r_2} - \frac{Y^2}{2r_1}$$

Hence

$$(D - d) = \frac{Y^2}{2}\left(\frac{1}{r_2} - \frac{1}{r_1}\right) = -\frac{Y^2}{2f'(n-1)}$$

$$(D - d)\,\Delta n = -\frac{Y^2}{2f'}\left(\frac{\Delta n}{n-1}\right) = -\frac{Y^2}{2f'V}$$

Now, by Eq. (49),

$$\text{paraxial chromatic aberration} = \frac{2\sum}{u'^2} = -\frac{Y^2}{f'Vu'^2}$$

in exact agreement with Eq. (39).

CHAPTER 5

Spherical Aberration

The direct calculation of spherical aberration is a simple matter. A meridional ray is traced from object to image, passing through the desired zone of a lens, and the image distance L is found. This is compared directly with the l' of a corresponding paraxial ray from the same object point. Then

$$\text{longitudinal spherical aberration} = LA' = L' - l'$$

I. SURFACE CONTRIBUTION FORMULAS

The simple relationship just given is often inadequate, both because it gives the aberration as a small difference between two large numbers, and also because it gives no clue as to where the aberration arises. It is therefore much more useful to compute the aberration as the sum of a series of surface contributions. A convenient formula has been given by Delano;[1] the derivation follows from Fig. 53. In this diagram, an entering marginal and paraxial

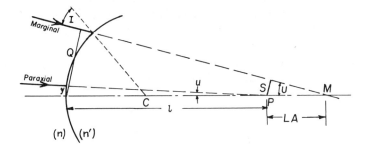

FIG. 53. Spherical aberration contribution.

ray are shown at a spherical surface. The length S is the perpendicular drawn from the paraxial object point P onto the marginal ray. The marginal ray is defined by its Q and U, the paraxial ray by its y and u. Then

$$S = Q - l \sin U, \qquad \text{hence} \qquad Su = Qu - y \sin U$$

[1] E. Delano, A general contribution formula for tangential rays, *J. Opt. Soc. Am.*, **42**, 631 (1952).

We now replace u on the right by $yc - i$ and $\sin U$ by $Qc - \sin I$, where c is the surface curvature as usual. Multiplying through by n gives

$$Snu = yn \sin I - Qni$$

Doing the same thing for the refracted ray and subtracting plain from prime gives

$$S'n'u' - Snu = (Q - Q')ni$$

We write this for every surface and add. After extensive cancellation because $(S'n'u')_1 = (Snu)_2$, we get for k surfaces,

$$(S'n'u')_k - (Snu)_1 = \sum (Q - Q')ni \tag{50}$$

Inspection of Fig. 53 shows that

$$LA = S/\sin U \qquad \text{and} \qquad LA' = S'/\sin U'$$

Hence

$$LA' = LA\left(\frac{n_1 u_1 \sin U_1}{n'_k u'_k \sin U'_k}\right) + \sum \frac{(Q - Q')ni}{n'_k u'_k \sin U'_k} \tag{51}$$

The quantity under the summation sign is the contribution of each surface to the spherical aberration of this particular ray, and the first term is the transfer of the object aberration across the lens to the image space. It may be thought of as the contribution of the object to the final aberration.

As an example of the use of this formula, we will take the lens used on p. 28. A marginal ray entering this lens from infinity at height 2.0 has been traced on p. 28, and a corresponding paraxial ray on p. 41. The additional data required for use of Delano's formulas are given in Table VIII.

It will be noted that the sum of the aberration contributions agrees closely with the $L - l'$ aberration obtained directly from ray tracing:

$$L = 11.29390$$
$$l' = 11.28586$$
$$LA' = L - l' = 0.00804$$

However, the L and l' values are good only to about 1 in the fifth decimal place whereas the contributions are good to 1 in the seventh place. The contribution method is clearly the more precise of the two.

Note, too, that the first and third surfaces of this lens contribute undercorrected aberration, the third giving twice as much as the first in spite of its flat curvature; the second surface contributes more overcorrection than the total undercorrection of the two outer surfaces in spite of the small index difference between the media on each side of it.

TABLE VIII

SURFACE CONTRIBUTIONS TO SPHERICAL ABERRATION

c	0.1353271		-0.1931098		-0.0616427	
d		1.05		0.40		
n		1.517		1.649		
		Paraxial ray data				
u	0	0.0922401		0.0554372		0.1666664
$yc - u = i$	0.2706542		-0.4597566		-0.1713855	
		Marginal ray data				
Q	2.0		1.9178334		1.9186619	
Q'	2.0171179		1.9398944		1.8814033	
$Q - Q'$	-0.0171179		-0.0220610		0.0372586	
ni	0.2706542		-0.6974508		-0.2826147	
$n'_k u'_k \sin U'_k$	0.0277643		0.0277643		0.0277643	
Spherical contribution	-0.1668701		0.5541815		-0.3792578	$\sum = 0.0080536$

An alternative representation of the contribution formula is sometimes useful. Its derivation depends on the relation between Q and the chord PA

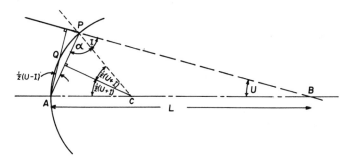

FIG. 54. Diagram showing that $Q = PA \cos \frac{1}{2}(U - I)$.

(Fig. 54). In triangle APB we have

$$\frac{PA}{\sin U} = \frac{L}{\sin(\alpha + I)}$$

Therefore,

$$PA = \frac{L \sin U}{\sin(\alpha + I)} = \frac{Q}{\sin(\alpha + I)}$$

However,

$$\alpha = 90° - \tfrac{1}{2}(U + I)$$

Therefore,

$$\alpha + I = 90° - \tfrac{1}{2}(U - I) \qquad \text{and} \qquad Q = PA \cos \tfrac{1}{2}(U - I)$$

Hence

$$(Q - Q') = PA[\cos \tfrac{1}{2}(U - I) - \cos \tfrac{1}{2}(U' - I')]$$
$$= PA[-2 \sin(\tfrac{1}{2} \text{ sum}) \sin(\tfrac{1}{2} \text{ diff})]$$
$$= 2PA \sin \tfrac{1}{2}(I' - U) \sin \tfrac{1}{2}(I' - I)$$

The spherical aberration contribution formula can therefore be written

$$LA' = LA\left(\frac{n_1 u_1 \sin U_1}{n'_k u'_k \sin U'_k}\right) + \sum \frac{2PA \sin \tfrac{1}{2}(I' - I) \sin \tfrac{1}{2}(I' - U)ni}{n'_k u'_k \sin U'_k} \qquad (52)$$

where $I' - I = U - U'$, and $I' - U = I - U'$.

A. THE THREE CASES OF ZERO ABERRATION AT A SURFACE

In Eq. (52), the quantity under the summation sign becomes zero in the following special cases:

(a) if $PA = 0$,
(b) if $I' = I$,
(c) if $i = 0$,
(d) if $I' = U$.

In case (a) the object and image are both at the vertex of the surface. In case (b) the marginal ray suffers no refraction at the surface; this could occur because the object is at the center of curvature of the surface, as also in case (c), but it could occur trivially if the refractive index were the same on both sides of the surface. Case (d) arises if $I' = U$ or if $I = U'$. This very important case must be considered further.

By Eq. (4a) we see that in this case

$$\sin I = Qc - \sin U = \left(\frac{L \sin U}{r}\right) - \sin U = \left(\frac{L}{r} - 1\right) \sin U$$

But $\sin I = (n'/n) \sin I'$, and since, in this special case, $I' = U$, we find that $(L/r) - 1 = n'/n$, whence

$$L = r(n + n')/n$$

and similarly

$$L' = r(n + n')/n'$$

It can also be shown that, for this particular pair of conjugates,

$$Q = Q', \qquad nL = n'L', \qquad 1/L + 1/L' = 1/r$$

We can understand this better with a numerical example. For a convex surface with air on the left and glass of index 1.5 on the right, we find that

$$L = 2.5r \qquad \text{and} \qquad L' = 1.6667r$$

These points are shown in Fig. 55. All rays in the object space directed

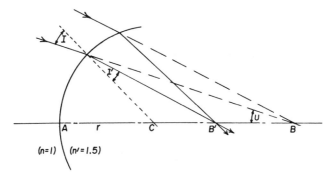

FIG. 55. The aplanatic points of a surface.

toward B will pass through B' after refraction, no matter at what angle they enter the surface. This pair of conjugates is known as the *aplanatic points* of the surface. Note that the distances of these points from the center of curvature of the surface are respectively equal to $r(n/n')$ and $r(n'/n)$. Aplanatic surfaces of this type are used in many types of lens, particularly high-power microscope objectives.

It must be borne in mind that an aplanatic surface is capable only of increasing the convergence of converging light or increasing the divergence of diverging light. The greater the convergence or divergence the greater will be the effect of an aplanatic surface. For parallel entering light, the aplanatic surface is a plane and produces no change in convergence.

B. AN APLANATIC SINGLE ELEMENT

It is possible to make an aplanatic single-element lens for use in a converging light beam by making the front face aplanatic and the rear face perpendicular to the marginal ray. Such a lens increases the convergence of a converging beam, which is useful in certain situations. In parallel light an aplanatic lens is merely a parallel plate. In a diverging beam an aplanatic lens element is a negative meniscus that increases the divergence of the beam without, of course, introducing any spherical aberration.

C. Effect of Object Distance on the Spherical Aberration Arising at a Surface

We have seen that the contribution of a single surface to the spherical aberration is zero if the (virtual) object is at *A*, *C*, or *B* in Fig. 55. We may now inquire what will happen if the object lies in any of the regions between these points, the light entering from the left in all cases. As an example, consider the case of a surface of radius 10 with air on the left and glass of index 1.5 to the right of it. We will let a ray enter this surface at a fixed slope angle of 11.5° (chosen because its sine is 0.2), and we calculate the spherical aberration in the image as the object moves along the axis. This is shown in Fig. 56.

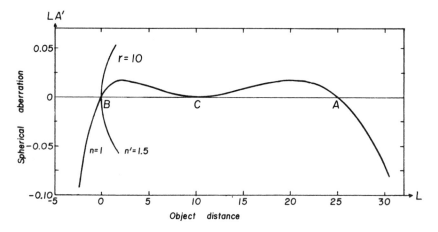

FIG. 56. Effect of object distance on spherical aberration.

If the object lies between the surface and the aplanatic point, a collective surface such as we are considering here contributes *overcorrected* spherical aberration, which is decidedly unexpected and can be quite useful. The peak value of this overcorrection occurs, in our case, when the object distance is about twice the surface radius. As can be seen in Fig. 56 the peak of overcorrection is close to the aplanatic point, and to achieve it we must use a surface somewhat flatter than the aplanatic radius. There is also a second but much less useful peak close to the surface itself, the value of L in this case being about 0.2 times the surface radius. As a general rule, using ordinary glasses, the maximum overcorrection will be obtained if r is set at a value about 1.2 times the aplanatic radius of $L/(1 + n')$, assuming that there is air to the left of the surface.

D. Effect of Lens Bending

One of the best methods for changing the spherical aberration of a lens is to "bend" it (see p. 56). If the lens is thin, changing all the surface curvatures by the same amount has the effect of changing the lens shape while leaving the focal length and the chromatic aberration unchanged. Generally spherical aberration varies with bending in a parabolic fashion when plotted against some reasonable shape parameter such as c_1. At extreme bendings either to left or right, a thin positive lens is decidedly undercorrected, and the aberration reaches a mathematical maximum at some intermediate bending. The aberration of a single thin lens with a distant object is never zero, but in a positive achromat the aberration exhibits a region of overcorrection at and close to the maximum. To bend a thick lens, it is customary to change all the surface curvatures except the last by a chosen value of Δc, the last radius being then solved by the ordinary angle solve procedure (p. 43) to maintain the paraxial focal length. This procedure, of course, slightly affects the chromatic aberration but it alters the spherical aberration far more. It should be noted, however, that if the aberration is at the maximum, then quite a significant bending will have little or no effect on the spherical aberration. Nevertheless, even then, bending can be used to vary other aberrations such as coma or field curvature.

E. A Single Lens Having Minimum Spherical Aberration

A single positive lens can be made to have minimum spherical aberration at one wavelength by taking a series of bendings, in each case solving for the last radius to hold focal length. When this is done, it is found that in the minimum-aberration lens each surface contributes about the same amount of aberration, with the front surface (in parallel light) contributing slightly more than the rear surface.

As an example, suppose we wish to design such a lens of focal length 10 and aperture $f/4$, using glass of index 1.523. A suitable thickness is 0.25. The results of ray tracing with $Y_1 = 1.25$ are as shown in the following tabulation,

c_1	0.15	0.16	0.17	0.18
Solved c_2	−0.041742	−0.031639	−0.021519	−0.011380
Spherical aberration contribution (1)	−0.05977	−0.07267	−0.08730	−0.10376
Spherical aberration contribution (2)	−0.10434	−0.08705	−0.07174	−0.05828
Total	−0.16411	−0.15972	−0.15904	−0.16205

The two surface contributions become equal at $c_1 = 0.1648$, for which bending the total aberration is found to be −0.15893. A careful plot shows that the true minimum occurs at $c_1 = 0.1670$ with an aberration of −0.15883, but of course the difference between these two values of the aberration is utterly insignificant. We shall therefore make no noticeable error by aiming at equal contributions for the minimum bending. The error may become much greater, however, in lenses made from high-index materials for the infrared.

F. A TWO-LENS MINIMUM ABERRATION SYSTEM

A considerable reduction in spherical aberration can be achieved by taking two identical lenses of twice the desired focal length and mounting them close together. In our case this procedure, after scaling to a focal length of 10, gave a spherical aberration of −0.0788, about half that of the original single element. However, a much greater improvement can be made by bending the second lens so that each of the four refracting surfaces contributes an identical amount of aberration. The required condition is that each surface should have the same value of $(Q - Q')ni$, since it is this product that determines the aberration contribution of the surface. The curvature of each surface is determined by a few trials, and then, if the resulting focal length is not correct, the whole lens is scaled up or down until it is.

As an example, suppose we add another element to the single minimum-aberration lens of Section E, finding c_3 and c_4 by trial to make all four contributions equal. This gives the lens shown in Fig. 57:

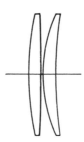

FIG. 57. A two-lens minimum aberration system.

c	d	n	Spherical aberration contribution for $Y_1 = 1.25$
0.1648			−0.01703
	0.25	1.523	
−0.02678			−0.01702
	0.05	(air)	
0.3434			−0.01703
	0.25	1.523	
0.1216			−0.01700

The focal length is now 4.6155 and the aperture is $f/1.85$. Scaling the lens to a focal length of 10.0 and tracing a ray at $Y_1 = 1.25$ ($f/4$) gives the total spherical aberration as −0.0310, about one-fifth that of the single element. It is worth noting that the focal lengths of the two elements are now not equal, being respectively 21.7 and 18.4.

There is a common misconception here, namely, that to secure minimum spherical aberration the marginal ray must be deviated equally at each of the four surfaces. To see how far this is from the truth, we list the surface ray deviations in the last example:

Surface	Angle U (deg)	Angle U' (deg)	Deviation $U' - U$ (deg)
1	0	1.877	1.877
2	1.877	3.318	1.441
3	3.318	6.019	2.701
4	6.019	7.195	1.176

The reason why the third surface does so much refracting " work " without the introduction of excessive aberration is its close proximity to the aplanatic condition.

It should be noted that when designing a two-element infrared lens with a material having a refractive index higher than about 2.5, such as silicon or germanium, it will be found that if r_3 is chosen to give the maximum possible overcorrection, it may actually overcompensate the undercorrection of the front minimum-aberrations lens, making it possible to correct the spherical aberration completely. The last radius is then chosen to have its center of curvature at the final image to eliminate any aberration there.

As an example, we will design an $f/1$ lens made of silicon having a refractive index of 3.4. Following the suggested procedure, we come up with:

c	d	n	Spherical aberration contribution	
0.02790			-0.006017	$f' = 10.283$
	0.25	3.4		
0.01572			-0.006004	$l' = 9.717$
	0.05	(air)		
0.12632			$+0.012009$	aperture $= 10$ ($f/1$)
	0.50	3.4		
0.10291			0	

Focal length of front component, 33.99; of rear component, 14.88

A longitudinal section of this lens is shown in Fig. 58. The strong rear

FG. 58. An $f/1$ silicon lens.

element is highly meniscus, as can be seen. High-index materials such as silicon and germanium appear to behave quite oddly to anyone familiar with the properties of ordinary glass lenses.

G. A Four-Lens Monochromat Objective

As has been stated, a single aplanatic lens element for use in parallel light is nothing but a planoparallel plate and not a lens at all. However, by making use of the small overcorrection that can be obtained from a convex surface slightly weaker than a true aplanat, it is possible to construct an aplanatic system for use with a distant object by placing a minimum-aberration lens first, and following this by a series of overcorrected menisci in the converging beam produced by the first lens.

As an example we may take the single minimum-aberration $f/4$ lens of p. 108, and follow it with three menisci, the front face of each being chosen to give the maximum of overcorrection, while the rear faces are perpendicular to the marginal ray. Nothing is gained by departing from the strict perpendicular condition for the rear surfaces of the menisci because, being dispersive surfaces, any departure from perpendicularity in either direction would yield spherical undercorrection, which is just what we are trying to avoid.

After several trials to obtain the greatest possible amount of overcorrection, and finally scaling to $f' = 10.0$ with an aperture of $f/2$, we obtain the following system:

c	d	n	Spherical aberration contribution at $f/2$
0.066014[a]	0.3	1.523	$\left.\begin{array}{l}-0.020622\\-0.020610\end{array}\right\}$ -0.041232
−0.010636[a]			
	0.05	(air)	
0.082192			+0.002463
	0.3	1.523	
0.055672			0
	0.05	(air)	
0.113932			+0.005962
	0.3	1.523	
0.077543			0
	0.05	(air)	
0.158867			+0.014476
	0.3	1.523	
0.109134			0
		Total	$\overline{-0.018331}$

[a] "Crossed" lens in parallel light (see p. 118).

The focal length of the first lens alone is now 24.969. It is clear that even with three menisci it is not possible to compensate for the undercorrection of the first lens.

However, we can do much better by starting with the two-lens minimum aberration form given on p. 109, and following this with only two menisci. By this procedure we can design a four-lens spherically corrected system for use in parallel light with an aperture as high as $f/2$. Scaled to $f' = 10$ this becomes

c	d	n	Spherical aberration contribution at $f/2$
0.041520			−0.005090
	0.3	1.523	
−0.006726			−0.005098
	0.05		
0.084883			−0.005106
	0.3	1.523	

Cont'd

Cont'd

c	d	n	Spherical aberration contribution at $f/2$
0.029164			−0.005098
	0.05		
0.113764			+0.005966
	0.3	1.523	
0.077891			0
	0.05		
0.159353			+0.014387
	0.3	1.523	
0.109941			−0.000016
		Total	−0.000068

This lens is shown in Fig. 59. The focal length of the first two lenses is now

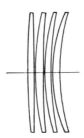

FIG. 59. A four-lens $f/2$ aplanatic objective.

18.380. This system has been used in monochromat microscope objectives made of quartz for use at a single wavelength in the ultraviolet. The design has been discussed by Fulcher.[2]

H. AN ASPHERIC PLANOCONVEX LENS FREE FROM SPHERICAL ABERRATION

Two cases arise, the first when the curved aspheric surface faces the distant object, and the other when the plane surface faces the object.

1. *Convex to the Front*

The image is considered to fall inside the glass, and the situation is indicated in Fig. 60a. By equating optical paths between any finite ray and

---

[2] G. S. Fulcher, Telescope objective without spherical aberration for large apertures, consisting of four crown glass lenses, *J. Opt. Soc. Am.* **37**, 47 (1947).

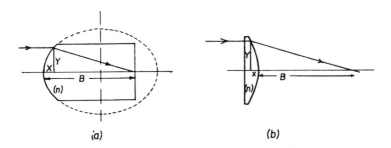

FIG. 60. Aspheric single lenses corrected for spherical aberration.

the axis, we find

$$Bn = X + n[(B - X)^2 + X^2]^{1/2}$$

whence

$$\frac{[X - Bn/(n + 1)]^2}{[Bn/(n + 1)]^2} + \frac{Y^2}{B^2(n - 1)/(n + 1)} = 1$$

This is an ellipse with semimajor axis a equal to $Bn/(n + 1)$ and semiminor axis b equal to $B[(n - 1)/(n + 1)]^{1/2}$; the eccentricity $e = [1 - (b/a)^2]^{1/2} = 1/n$. For example, if $B = 20$ mm and $n = 1.5$, the semiaxes become 12.0 and 8.94, respectively, and the vertex radius is 6.66 with $e = 0.6666$. A surface of this kind has long been used on highway reflector "buttons."

2. *Plane Surface in Front*

Equating optical paths in the air behind the lens gives

$$B + nX = [Y^2 + (B + X)^2]^{1/2}$$

whence

$$\frac{\{X + [B/(n + 1)]\}^2}{[B/(n + 1)]^2} - \frac{Y^2}{B^2(n - 1)/(n + 1)} = 1$$

There is a clear resemblance between these two cases. The plane-in-front lens has a hyperbolic surface with semimajor axis equal to $B/(n + 1)$, and semiminor axis equal to $B[(n - 1)/(n + 1)]^{1/2}$ as before (Fig. 60b), the eccentricity now being equal to the refractive index n. This surface can be applied on both faces of a biconvex lens for use at finite magnification, up to as high an aperture as required, without any spherical aberration.

II. ZONAL SPHERICAL ABERRATION

As we have seen, it is possible by the use of opposing positive and negative elements to design a lens such that the focus of the marginal ray coincides with the paraxial image point. We say that this lens has zero spherical aberration. However, it generally happens that the foci of rays passing through the intermediate zones of the lens fall closer to the lens than the paraxial image-point, and occasionally but rarely fall further from it. Thus we can plot a graph connecting entrance height Y with the spherical aberration, as in Fig. 61. This zonal residual is generally known as *zonal aberration*. It can be expressed as a power series containing only even powers of Y, as

$$LA' = aY^2 + bY^4 + cY^6 + \cdots$$

The successive terms of this series have been called primary, secondary, . . . aberration, but of course they have no separate existence, and the actual aberration of the lens is the sum of all these terms. However, we can plot them separately to see how they vary (Fig. 61). If Y is small, the secondary

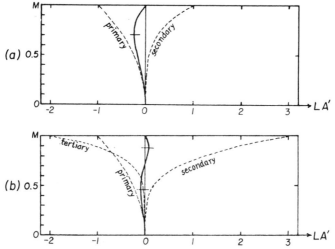

FIG. 61. Interaction of various orders of spherical aberration. (a) Primary and secondary only; (b) primary, secondary, and tertiary.

and higher terms are very small or negligible, and the primary term represents the whole aberration. Then, at increasing values of Y, first the secondary and then the tertiary terms begin to increase and finally dominate the situation. In the example shown in Fig. 61a, the primary term is negative and the secondary term is positive, and they have equal and opposite values for the marginal ray. When only the first two terms are present, as in this example, the peak zonal residual occurs when Y is equal to the marginal Y_m multiplied by $1/\sqrt{2} = 0.7071$. The magnitude of the zonal residual is, in this case, equal to one-quarter of the primary term at the marginal zone of the lens.

Because tertiary aberration is not greatly different from secondary, it may be positive and add to the secondary; in this case the maximum zonal residual falls higher than the 0.7 zone, and the marginal aberration increases very rapidly. On the other hand, if the tertiary aberration is negative, it tends to oppose the secondary, and it is then possible to eliminate both the marginal and the zonal aberration, as indicated in Fig. 61b. It will be noticed that the secondary and tertiary aberrations are now much larger than in the simple case of Fig. 61a, but the resulting aberration curve is nearly flat, having small equal and opposite residuals above and below the 0.7 zone. An analysis of the situation reveals that the maximum and minimum residuals fall at values of Y given by

$$Y/Y_m = (1 \pm 1/\sqrt{3})^{1/2}/\sqrt{2} = 0.8881 \quad \text{or} \quad 0.4597$$

The locations of the maximum residuals are indicated by short horizontal lines on these diagrams.

III. PRIMARY SPHERICAL ABERRATION

A. AT A SINGLE SURFACE

To isolate the primary term, we would have to make Y of infinitesimal magnitude, and then we cannot use the formula in Eq. (51) to compute the aberration for the same reason that we cannot trace a paraxial ray by the ordinary ray-tracing formulas. However, the primary term can be determined as a limit:

$$LA'_{\text{primary}} = \lim_{y \to 0} (LA'_y)$$

To find this limit, we use paraxial ray data to fill in the numbers in the accurate Eq. (52). Making this substitution gives the primary aberration equation:

$$LA'_p = LA_p\left(\frac{n_1 u_1^2}{n'_k u'^2_k}\right) + \sum \frac{2y \cdot \frac{1}{2}(i' - i) \cdot \frac{1}{2}(i' - u)ni}{n'_k u'^2_k}$$

Here LA_p is the primary aberration of the object, if any; it is transferred to the final image by the ordinary longitudinal magnification rule. The quantity under the summation sign is the primary aberration arising at each surface.

These surface contributions can be written

$$SC = yni(u - u')(i - u')/2n'_k u'^2_k \qquad (53)$$

Only paraxial ray data are required to evaluate this formula. To interpret it, we note that for pure primary aberration,

$$LA'_p = aY^2$$

and that the radius of curvature of the spherical-aberration graphs in Fig. 61, at the point where the graph crosses the axis, has the value

$$\rho = Y^2/2LA'_p = 1/2a$$

Therefore, the coefficient of primary aberration a is an inverse measure of twice the radius of curvature of the spherical-aberration graph at the point where it crosses the lens axis. Hence, by tracing one paraxial ray, we not only discover the location of the image point, but we also asertain the shape of the aberration curve as it crosses the axis at that point. It is remarkable how much information can be obtained from so very little ray-tracing effort.

As an example of the use of this formula, we will calculate the primary spherical aberration contributions of the three surfaces of the cemented doublet on p. 28 that we have already used several times (see Table IX). It is interesting to compare these primary aberration contributions with the exact contributions given on p. 103. The contributions are

Surface	1	2	3	Sum
Exact contribution	−0.16687	+0.55418	−0.37926	0.00805
Primary contribution	−0.16035	+0.45301	−0.35979	−0.06713
Difference (higher orders)	−0.00652	+0.10117	−0.01947	

At each surface the true and primary contributions are similar in magnitude and have the same sign, but the cemented interface shows the greatest difference. This is due to the presence of a significant amount of secondary and higher aberrations there, while the outer surfaces show very little sign of higher aberrations. It is the presence of the considerable amount of higher aberrations at the cemented interface that is the cause of the large zonal aberration of this lens.

B. PRIMARY SPHERICAL ABERRATION OF A THIN LENS

By combining the SC values for the two surfaces of a thin lens element, we find that a thin lens, or a thin group of lenses in close contact, within a

TABLE IX

CALCULATION OF PRIMARY SPHERICAL ABERRATION

y	2	1.9031479	1.8809730	
n	1	1.517	1.649	
$yc - u = i$	0.2706542	-0.4597566	-0.1713855	
u	0	0.0922401	0.0554372	0.1666664
y	2	1.9031479	1.8809730	
ni	0.2706542	-0.6974508	-0.2826147	
$u - u'$	-0.0922401	0.0368029	-0.1112292	
$i - u'$	0.1784141	-0.5151938	-0.3380519	
$1/2u_k'^2$	18	18	18	
Product $= SC$	-0.160349	0.453014	-0.359792	$\sum = -0.067127$

system contributes the following amount to the primary spherical aberration at the final image:

$$SC = -\frac{y^4}{n_0' u_0'^2} \sum (G_1 c^3 - G_2 c^2 c_1 + G_3 c^2 v_1 + G_4 cc_1^2 - G_5 cc_1 v_1 + G_6 cv_1^2)$$

$$(54)$$

where the terms with suffix 0 refer to the final image, the other terms applying to each single element. Here c and c_1 have their usual meanings, namely, $c_1 = 1/r_1$ and $c = 1/f'(n - 1)$. The symbol v_1 is the reciprocal of the object distance of the element, and the G are functions of the refractive index, namely,

$$G_1 = \tfrac{1}{2}n^2(n - 1), \qquad G_2 = \tfrac{1}{2}(2n + 1)(n - 1),$$

$$G_3 = \tfrac{1}{2}(3n + 1)(n - 1), \qquad G_4 = \tfrac{1}{2}(n + 2)(n - 1)/n,$$

$$G_5 = 2(n^2 - 1)/n, \qquad G_6 = \tfrac{1}{2}(3n + 2)(n - 1)/n$$

The details of the derivation of this formula have been given by Conrady.[3] The summation sign in expression (54) is used only if the thin lens contains more than one element, e.g., if it is a thin doublet or triplet, otherwise it may be omitted. If there is more than one element we must assume a very thin layer of air to exist between the elements in place of cement, c_1 being the curvature of the first surface of each element and v_1 being the reciprocal of the object distance in air. Thus for the second lens of a cemented doublet we take

$$(c_1)_b = (c_1)_a - c_a \qquad \text{and} \qquad (v_1)_b = (v_1)_a + c_a(n_a - 1)$$

---

[3] A. E. Conrady, p. 95.

In the case of an isolated thin element or thin system in air, not forming part of a more complex system, $n'_0 = 1$ and $u'_0 = y/l'$. Also the aberration of the object (if any) must be transferred to the image and added to the new aberration arising at the lens. Thus in such a case we have

$$LA'_p = LA_p \left(\frac{l'}{l}\right)^2 - y^2 l'^2 \sum (G \text{ sum})$$

the (G-sum) referring to the six-term expression in parentheses in Eq. (54).

By use of the G-sum formula we can plot a graph showing how the primary spherical aberration of a thin lens varies with the bending. (Fig. 62).

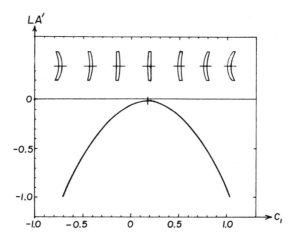

FIG. 62. Effect of bending on spherical aberration.

For a single thin positive element this graph is a vertical parabola, the vertex of which nearly but not quite reaches the zero aberration line.

The thin single lens having the minimum primary spherical aberration is called a *crossed lens*; its shape can be found by differentiating the G-sum expression with respect to c_1:

$$dLA'/dc_1 = -y^2 l'^2 (-G_2 c^2 + 2G_4 cc_1 - G_5 cv_1) = 0$$

whence the value of c_1 corresponding to the crossed lens is given by

$$c_1 = (G_2 c + G_5 v_1)/2G_4$$

Inserting the explicit values of the G functions gives

$$c_1 = \frac{\frac{1}{2}n(2n+1)c + 2(n+1)v_1}{n+2} \tag{55}$$

For the special case of a very distant object, $v_1 = 0$, and then

$$c_1/c = n(2n + 1)/2(n + 2)$$

whence

$$c_2/c_1 = (2n^2 - n - 4)/n(2n + 1) = r_1/r_2$$

For a series of typical refractive indices, this formula gives

n:	1.5	1.6	1.7	2.0	3.0	4.0
c_2/c_1:	-6	-14	$+94$	$+5$	$+1.9$	$+1.5$

For glass having an index of 1.6861, the crossed lens is exactly planoconvex, but for other glass indices the departure from the planconvex form is slight. The very high indices 3.0 and 4.0 refer, of course, to infrared materials, and here the crossed lens is found to be a deeply curved thin meniscus.

IV. THE IMAGE DISPLACEMENT CAUSED BY A PLANOPARALLEL PLATE

From Fig. 63 it is clear that the longitudinal image displacement caused

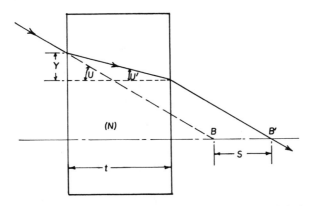

FIG. 63. Image displacement caused by the insertion of a parallel plate.

by the insertion of a thick planoparallel plate into the path of a ray having a convergence angle U is $S = BB'$, given by

$$S = \frac{Y}{\tan U'} - \frac{Y}{\tan U}$$

$$= \frac{Y}{\tan U'}\left(1 - \frac{\tan U'}{\tan U}\right)$$

But $Y/\tan U'$ is equal to t, the thickness of the plate. Therefore,

$$S = t\left(1 - \frac{\tan U'}{\tan U}\right) = \frac{t}{N}\left(N - \frac{\cos U}{\cos U'}\right)$$

where N is the refractive index of the plate. For a paraxial ray this reduces to

$$s = \frac{t}{N}(N - 1)$$

The plate of glass occupies more space than its "air equivalent," which is defined as that thickness of air in which a paraxial ray drops or rises by the same amount as in the glass plate. Thus

air equivalent = glass thickness/refractive index

To the lens designer a reflecting prism in a system behaves as though it were a very thick parallel plate. In a converging beam a prism has the effect of overcorrecting the three longitudinal aberrations, spherical, chromatic, and astigmatism, while it undercorrects the transverse aberrations, coma, distortion, and lateral color.

V. SPHERICAL ABERRATION TOLERANCES

A. PRIMARY ABERRATION

Conrady has shown[4] that if a lens suffers from a small amount of pure primary spherical aberration, the best-fitting reference sphere will touch the emerging wavefront at center and edge, and the plane of best focus will lie midway between the marginal and paraxial image points. If the aberration is large compared with the Rayleigh limit, geometrical considerations dominate, and the geometrical circle of least confusion becomes the "best" focus. However, there is also a secondary best focus close to the paraxial focus; this has been amply verified by experiment.[5]

In the case of pure primary aberration, the magnitude of the maximum residual OPD at this best focus is equal to the Rayleigh quarter-wave limit when

$$LA' = 4\lambda/\sin^2 U'_m = 16\lambda(f \text{ number})^2 \tag{56}$$

where f number = focal length/diameter of aperture.

---

[4] A. E. Conrady, p. 628.

[5] H. G. Conrady, An experimental study of the effects of varying amounts of primary spherical aberration on the location and quality of optical images, *Phot. J.* **66**, 9 (1926).

This aberration tolerance is surprisingly large, being four times the extent of the focal range. Some typical values for $\lambda = 0.0005$ mm are given in the following tabulation

f Number	4.5	6	8	11	16	22
Primary aberration tolerance (mm)	0.2	0.3	0.5	1.0	2.0	3.9

B. ZONAL ABERRATION

Conrady has also shown[6] that if a lens is spherically corrected for the marginal ray, the residual zonal aberration will reach the Rayleigh limit if its magnitude is

$$LZA = 6\lambda/\sin^2 U'_m \qquad (57)$$

or 1.5 times the tolerance for pure primary spherical aberration.

For telescopes, microscopes, projection lenses, and other visual systems, it is best not to allow any overcorrection of the marginal spherical aberration, even though this would reduce the zonal residual. This is because overcorrection leads to an unpleasant haziness of the image, and the zonal tolerance is so large that it is not likely to be exceeded. Indeed, many projection lenses are deliberately undercorrected, even for the marginal ray, to give the cleanest possible image with maximum contrast. Photographic objectives, on the other hand, are generally given an amount of spherical overcorrection equal to two or three times the zonal undercorrection. The overcorrected haze is often too faint to be recorded on film, especially if the exposure is on the short side, and in any case the lens will generally be stopped down somewhat, which cuts off the marginal overcorrection leaving a small and often quite insignificant zonal residual.

In this connection, it may well be pointed out that focusing a camera by unscrewing the front element has the effect of rapidly undercorrecting the spherical aberration. This leads to a loss of definition and some degree of focus shift at small apertures, but its convenience to the camera designer outweighs these objections. If the lens is known to be intended for this type of focusing, then it should be designed with a large amount of spherical overcorrection. If possible, the aberration should be well corrected at a focus distance of about 15–20 feet.

C. CONRADY'S OPD'_m FORMULA

Probably the best way to ascertain if a lens is adequately corrected for

---

[6] A. E. Conrady, p. 631.

zonal aberration is to calculate the optical path difference between the emerging wavefront and a reference sphere centered about the marginal image point. Conrady[7] has given a formula by which the contribution of each lens surface to this OPD can be found:

$$OPD'_m = \frac{Yn \sin I \sin \frac{1}{2}(U - U') \sin \frac{1}{2}(I - I')}{2 \cos \frac{1}{2}(U + I) \cos \frac{1}{2}U \cos \frac{1}{2}I \cos \frac{1}{2}U' \cos \frac{1}{2}I'} \tag{58a}$$

Referring back to Eq. (52), we see that by using the Q method of ray tracing, Conrady's expression can be greatly simplified to

$$OPD'_m = \frac{(Q - Q')n \sin I}{4 \cos \frac{1}{2}U \cos \frac{1}{2}I \cos \frac{1}{2}U' \cos \frac{1}{2}I'} \tag{58b}$$

This OPD term has the same sign as the spherical aberration contribution at any surface. If the lens is spherically corrected for the marginal ray, the magnitude of this sum is a measure of the zonal aberration, the sum being positive for a negative zone. The advantage of using the OPD formula is that the tolerance of the sum is known to be two wavelengths. Hence we have an immediate assessment of the significance of the zonal residual; this is much more accurate than the simple zonal tolerance given on p. 121, which is valid only for a mixture of primary and secondary aberrations.

If the spherical aberration is zero at both margin and 0.7 zone, as in the diagram of Fig. 61b, then we can determine the seriousness of the two remaining small zones by calculating the OPD sum along the marginal ray (which should be zero) and also along the 0.7 zonal ray.

---

[7] A. E. Conrady, p. 616.

CHAPTER 6

Design of a Spherically Corrected Achromat

Since the chromatic aberration of a lens depends only on its power, whereas the spherical aberration varies with bending, it is obviously possible to select that bending of an achromat that will give us any desired spherical aberration (with limits). There are two possible approaches to this design. The first is the four-ray method, requiring no optical knowledge, and the second makes use of a thin-lens study based on primary aberration theory to guide us directly to the desired solution. The latter method is by far the most desirable since it also indicates how many possible solutions there are to any given problem.

I. THE FOUR-RAY METHOD

In this procedure we set up a likely first form, which can actually be rather far from the final solution, and determine the spherical aberration by tracing a marginal ray and a paraxial ray in D light, and we calculate the chromatic aberration by tracing 0.7 zonal rays in F and C light. We then make trial changes in c_2 and c_3, keeping c_1 fixed, using a "double graph" to indicate what changes should be made to reach the desired solution. This simple but effective procedure is sometimes called the brute force method; it is especially convenient if a small computer is available for ray tracing.

As an example we will use this procedure to design an achromatic doublet with a focal length of 10 and an aperture 2.0 ($f/5$) using the following glasses:

	n_C	n_D	n_F	Δn	V
(a) Crown	1.52036	1.523	1.52929	0.00893	58.6
(b) Flint	1.61218	1.617	1.62904	0.01686	36.6
				$V_a - V_b = 22.0$	

The thin-lens (c_a, c_b) formulas on p. 79 for an achromat give

$$c_a = 0.5090, \qquad c_b = -0.2695$$

123

and if we assume that the crown element is equiconvex, our starting system will be

$$c_1 = 0.2545, \qquad c_2 = -0.2545, \qquad c_3 = 0.0150$$

By means of a scale drawing of this lens (setup A) we assign suitable thicknesses of 0.4 for the crown element and 0.16 for the flint. The result of ray tracing is

	$Y = 1$	$Y = 0.7$
	$L_D' = 9.429133$	$L_F' = 9.426103$
	$l_D' = 9.429716$	$L_C' = 9.430645$
spherical aberration = -0.000583		chromatic aberration = -0.004542

We next make a trial change in c_3 by 0.002 (setup B). This gives spherical aberration = $+0.001304$ and chromatic aberration = -0.001533. In addition, a further trial change in c_2 by 0.002 (setup C) gives spherical aberration = -0.002365 and chromatic aberration = -0.003027. These changes are plotted on a graph connecting chromatic aberration as ordinate with spherical aberration as abscissa (Fig. 64). Drawing a line

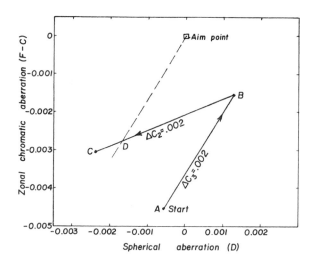

FIG. 64. The four-ray method for designing a cemented doublet.

through the aim point (0, 0) parallel to the line AB, intersecting line BC at D, suggests that we should try the following changes from setup B:

$\Delta c_2 = 0.00164$, but c_2 was -0.2545; therefore try $c_2 = -0.25286$,

$\Delta c_3 = 0.00181$, but c_3 was 0.0170; therefore try $c_3 = 0.01881$.

Ray tracing this system gives for the final setup:

c	d	n_D	V
0.2545			
	0.4	1.523	58.6
−0.25286			
	0.16	1.617	36.6
0.01881			

$f' = 10.0916$, $l' = 9.6288$, $LA'(f/5) = -0.00005$, and $L'_{ch} = +0.00004$.
Evidently the aberration changes are highly linear in this particular type of
lens. We shall find many applications of this "double-graphing" technique
whenever we are trying to correct two aberrations by making two simultaneous changes in the lens parameters.[1]

II. A THIN-LENS PREDESIGN

For the predesign of an ordinary cemented doublet, we start by determining the c_a and c_b values for thin-lens chromatic correction as described
on p. 79. We then set up the G-sum expressions for the primary spherical
aberration of a thin system as described on p. 117. Since we shall be using c_1
as a bending parameter, we express everything in terms of c_1. For the crown
element, c is c_a, c_1 remains as c_1, and v_1 is the reciprocal of the object
distance. For the flint element, $c_3 = c_1 - c_a$, since the two elements are to be
cemented together, c is c_b, and $v_3 = v_1 + (n_a - 1)c_a$. The sum of the two G
sums is now a quadratic in c_1, which can be solved either mathematically or
graphically to give the two values of c_1 that meet the requirements of the
problem. It can be seen that there are actually two and only two solutions;
the four-ray method gave only the solution closest to the arbitrary starting
setup and totally ignored the second solution.

As an example we will use glasses similar to those used for the four-ray
method, giving $c_a = 0.5085$ and $c_b = -0.2679$. For the G sums, with crown
lens in front, we have $f^2y^2 = 100$, $v_1 = 0$, $v_3 = 0.2659$, and $c_3 = c_1 - 0.5085$.

---

[1] This procedure was suggested to the author by his colleague, Mr. H. F. Bennett.

Using these values the spherical G sums give

$$SC_a = -30.759c_1^2 + 27.357c_1 - 7.9756$$
$$SC_b = 18.543c_1^2 - 23.698c_1 + 7.8392$$
$$\text{total} = -12.216c_1^2 + 3.659c_1 - 0.1364 \tag{59}$$

Evaluating this expression for a series of values of c_1 enables us to plot a graph of spherical aberration against c_1 (Fig. 65) from which our two pos-

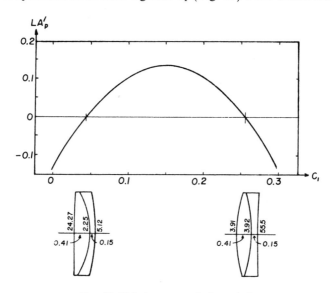

FIG. 65. Thin-lens crown-in-front designs.

sible solutions can be picked off. It should be reiterated that this graph is incorrect for three reasons: it assumes thin lenses, it considers only paraxial chromatic aberration, and it considers only primary spherical aberration. Nevertheless, the two solutions come out to be surprisingly close to the final solutions.

A. INSERTION OF THICKNESS

Since we require zero spherical aberration, we read off the two solutions as

$$c_1 = 0.044 \qquad \text{or} \qquad c_1 = 0.256$$

We now make a scale drawing of these systems and insert suitable thicknesses of 0.415 and 0.15, respectively. We next trace a marginal ray in D light and calculate the last radius for perfect achromatism by the $D - d$ method, as explained on p. 95. We complete the trace of the marginal ray and add a paraxial ray so that the true spherical aberration can be found. Since this will not be quite the desired value, although it will generally be very close, we find dLA'/dc_1 by differentiating Eq. (59), and apply the coefficient to ascertain how much the c_1 must be changed to eliminate the spherical aberration residual. The results for the two solutions are shown in the following tabulation:

c_1	0.044	0.256
Accurate LA'	0.0072	−0.0007
dLA'/dc_1	2.584	−2.596
Δc_1	−0.0028	−0.0003
New c_1	0.0412	0.2557
New LA'	−0.0001	0.0000
LZA'	−0.0171	−0.0045

The two final designs are as follows:

Left-hand solution			Right-hand solution		
c	d	n	c	d	n
0.0412			0.255755		
	0.412	1.523		0.415	1.523
−0.4442			−0.255037		
	0.15	1.617		0.15	1.617
−0.1953			0.018021		
$f' = 9.9943$			$f' = 9.99398$		
$l' = 9.9545$			$l' = 9.52719$		
$(f/5)\begin{cases} LA' = -0.00007 \\ LZA = -0.01705 \end{cases}$			$(f/5)\begin{cases} LA' = -0.0000 \\ LZA' = -0.00450 \end{cases}$		

Scale drawings of the two systems are included in Fig. 65. The decision as to which is the better design is based on the zonal aberration, which is nearly five times as large in the left-hand design as in the right-hand form. Furthermore, the surfaces in the right-hand design are weaker than in the left, resulting in economy in manufacture, and the fact that the crown element is almost equiconvex suggests that it should be made perfectly equiconvex to simplify the cementing operation. To do this requires a slight bending to the left, which would introduce a small spherical overcorrection, but it would probably be better to hold the spherical correction by varying the last radius, and accept the slight chromatic residual that would result. To complete the design, we calculate marginal, zonal, and paraxial rays in three wavelengths and plot the spherochromatism graph in Fig. 66.

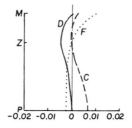

FIG. 66. Spherochromatism of right-hand $f/5$ crown-in-front solution.

B. FLINT-IN-FRONT SOLUTIONS

There is no magic about having the crown element in front, and indeed for some applications a flint-in-front form is preferred. Repeating the predesign procedure with the flint glass as a and the crown glass as b gives

$$\text{spherical aberration} = -12.2162c_1^2 + 5.6493c_1 - 0.5399 \quad (60)$$

This is plotted in Fig. 67, from which we see that the two spherically cor-

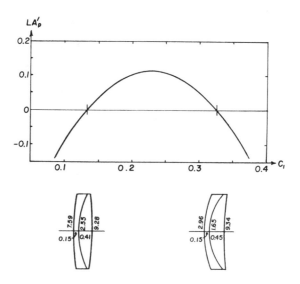

FIG. 67. Flint-in-front solutions.

rected forms are as follows:

c_1	0.135	0.327
Accurate LA'	0.0078	0.0242
dLA'/dc_1	2.351	-2.340
Δc_1	-0.0033	0.0103
New c_1	0.1317	0.3373
Accurate LA'	-0.0002	0.0004
Accurate LZA'	-0.0052	-0.0194

The two final flint-in-front designs are as follows:

Left-hand solution			Right-hand solution		
c	d	n	c	d	n
0.1317			0.3373		
	0.15	1.617		0.15	1.617
0.3917			0.6052		
	0.414	1.523		0.454	1.523
-0.1079			0.108114		

$$f' = 9.9963 \qquad f' = 10.0564$$
$$l' = 9.7994 \qquad l' = 9.4056$$
$$(f/5)\begin{cases}LA' = -0.00015\\ LZA = -0.0052\end{cases} \qquad (f/5)\begin{cases}LA' = 0.00037\\ LZA = -0.0194\end{cases}$$

Consideration of all four solutions indicates clearly that the right-hand crown-in-front form is in every way the best, although the zonal aberration of the left-hand flint-in-front form is not significantly greater. However, the weakness of the radii and the possibility of making the crown element exactly equiconvex are sufficiently important to render the crown-in-front form generally preferable.

C. FLOW CHART FOR A COMPUTER PROGRAM

The whole process of designing a cemented achromat can be programmed for a small computer. We have to enter into the program the following data: the n_a, V_a, n_b, and V_b of the two glasses; the focal length and its tolerance; the edge thickness of the positive element and the center thickness of the negative; the desired residual of the $D - d$ sum for chromatic aberration; the desired residual of spherical aberration and its tolerance; and a statement as to whether the right- or left-hand solution is required. The flow chart, Fig. 68, refers to a positive doublet with crown in front; in

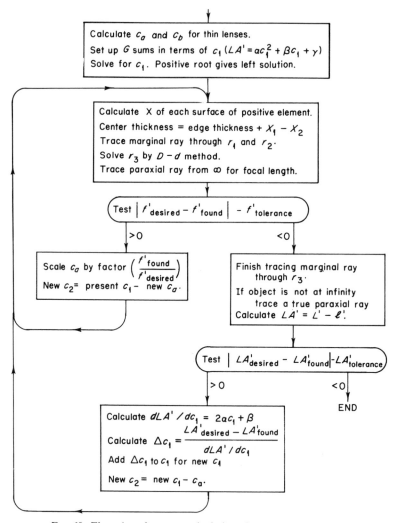

FIG. 68. Flow chart for automatic design of a cemented doublet.

practice it will have to be modified to include the possibility of flint-in-front forms and negative achromats. Some degree of control must also be applied to take care of the case where the required spherical aberration can never be reached, especially if a large amount of overcorrection is specified or an unsuitable glass choice has been made. Another situation in which the program may break down is at the top of the bending parabola where the derivative dLA'/dc_1 is zero, because then the calculated Δc_1 becomes infinite and quite meaningless.

III. CORRECTION OF ZONAL SPHERICAL ABERRATION

If the zonal aberration in a lens system is found to be excessive, it can often be reduced by splitting the system into two lenses each having half the lens power, in a manner analogous to the reduction of the marginal aberration of a single lens (see p. 108).

Another method that is frequently employed in a cemented system is to separate the cemented interface by a narrow parallel airgap. For this procedure to be effective, there must be a large amount of spherical aberration in the airgap so that the marginal ray drops disproportionately rapidly as compared to the 0.7 zonal ray. The airgap therefore undercorrects the marginal aberration more rapidly than the zonal aberration. As the rear negative element is now not acting as strongly as before because of the reduction of incidence height, the last radius must be adjusted to restore the chromatic correction, ordinarily by use of the $D - d$ method. As the spherical aberration will now be strongly undercorrected, it must be restored by a bending of the whole lens. Using this procedure, it is often possible to correct both the marginal and the zonal aberrations simultaneously.

To determine the proper values of the airgap and the lens bending, we start with a cemented lens and introduce an arbitrary small parallel airgap, the last radius being found by the $D - d$ method. The whole lens is then bent by trial until the marginal aberration is correct, and the zonal aberration is found. If it is still negative, a wider airgap is required. The desired values are quickly found by plotting suitable graphs.

As an example, we may consider the following three $f/3.3$ systems. They each have a focal length of 10.0, and they are made from K-3 and F-4 glasses, the last radius in each case being found by the usual $D - d$ procedure:

A			B			C		
c	d	n	c	d	n	c	d	n
0.259			0.259			0.236		
	0.75	1.51814		0.75	1.51814		0.75	1.51814
−0.2518			−0.2518			−0.2748		
	0.25	1.61644		0.0162	(air)		0.0162	(air)
0.018048			−0.2518			−0.2748		
				0.25	1.61644		0.25	1.61644
			0.022487			−0.005068		
$LA_{marg} = 0.001252$			−0.114863			−0.000211		
$LA_{zonal} = -0.024094$			−0.058951			0.000345		

System A is a well-corrected doublet of the ordinary type, but of unusually high aperture so as to illustrate the principle. The spherical aberration curve is shown in Fig. 69a. After introducing an airgap and suitably strengthening

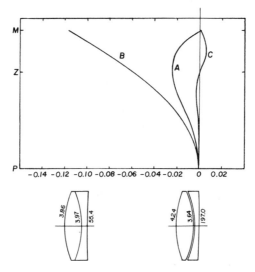

FIG. 69. Effect of a narrow airgap on spherical aberration. (a) Cemented doublet; (b) effect of introducing a narrow airgap; (c) final solution.

the last radius by the $D - d$ method, we have system B. The change in aberrations as a result of the introduction of this airgap is

$$\left.\begin{array}{l} \Delta LA_{marg} = -0.116115 \\ \Delta LA_{zonal} = -0.034857 \end{array}\right\} \quad \text{ratio } 3.33$$

We now bend the entire system to the left by $\Delta c = -0.023$ to restore the aberrations. The changes now are

$$\left.\begin{array}{l} \Delta LA_{marg} = 0.114652 \\ \Delta LA_{zonal} = 0.059296 \end{array}\right\} \quad \text{ratio } 1.93$$

If everything were ideal and only primary and secondary aberration were present, the latter ratio would be 2.0, and so we see that the changes due to bending are fairly linear in this respect.

Unfortunately, although the marginal and 0.7 zonal aberrations are virtually zero in system C, there are sizable intermediate zonal residuals remaining. By tracing a few additional zonal rays at various heights of incidence, we can plot the spherical aberration graph of this system (see Fig. 69c). However, it will be seen that these unavoidable residuals are much smaller than the 0.7 zonal aberration of the original cemented system A.

We can apply this same procedure to reduce zonal aberration by thickening a lens element, provided there is a large amount of undercorrected aberration within the glass. This is done frequently in photographic objectives, such as in double-Gauss lenses of high aperture.

IV. DESIGN OF AN APOCHROMATIC OBJECTIVE

A. A CEMENTED DOUBLET

A simple cemented doublet can be made apochromatic if suitable glasses are chosen in which the partial dispersion ratios are equal. The combination of fluorite and dense barium crown mentioned on p. 82 is one possibility. Another is a doublet made from two of the newer Schott glasses such as in the following tabulation:

Glass	n_e	$\Delta n = (n_F - n_C)$	$V_e = \left(\dfrac{n_e - 1}{n_F - n_C}\right)$	P_{Fe}
FK-52	1.48747	0.00594	82.07	0.4562
KzFS-2	1.56028	0.01036	54.08	0.4562

The large V difference of 27.99 keeps the elements weak and reduces the zonal aberration.

B. A TRIPLE APOCHROMAT

Historically the preferred form for an apochromatic telescope objective has been the apochromatic triplet or "photovisual" objective suggested by Taylor in 1892.[2] The preliminary thin-lens layout has already been described on p. 84, and we shall now proceed to insert thicknesses and find that bending of the lens which removes spherical aberration. The net curvatures and glass data of the thin system are given on p. 86. The glass indices and other data are stated to seven decimal places, by use of the interpolation formulas given in the current Schott catalog; this extra precision is necessary if the computed tertiary spectrum figures are to be meaningful. Obviously, in any practical system such precision could never be attained.

A possible first thin-lens setup with a focal length of 10 is the following:

$$c_1 = 0.56 \quad \text{(say)} \qquad r_1 = 1.79 \quad \text{(approx.)}$$

$$c_a = 1.0090432$$

$$c_2 = c_1 - c_a = -0.4490432 \qquad r_2 = -2.23$$

---

[2] H. D. Taylor, Br. Patent 17994/1892.

$$c_b = -0.7574313$$

$$c_3 = c_2 - c_b = 0.3083881 \qquad r_3 = 3.24$$

$$c_c = 0.1631915$$

$$c_4 = c_3 - c_c = 0.1451966 \qquad r_4 = 6.89$$

Tracing paraxial rays through this lens with all the thicknesses set at zero gives the image distances previously plotted in Fig. 45.

Since an aperture of $f/8$ is the absolute maximum for such a triple apochromat, we draw a diagram of this setup at a diameter of 1.25, by means of which we assign suitable thicknesses, respectively 0.3, 0.13, and 0.18. This lens is shown in Fig. 70a. Our next move is to trace a paraxial ray in e light

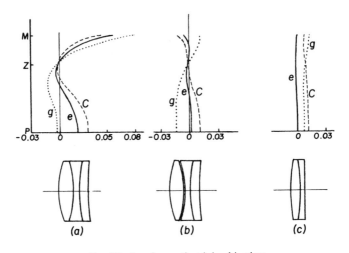

FIG. 70. Apochromatic triple objectives.

through this thick system, and as we go along modify each surface curvature in such a way as to restore the paraxial chromatic aberration contribution to its thin-lens value. Since the chromatic contribution is given by

$$L'_{ch} C = yni(\Delta n/n - \Delta n'/n')/u_k'^2 \qquad (37b)$$

it is clear that all we have to do is to maintain the value of the product (yi) at each surface. The equations to be solved, therefore, are

$$i = \frac{\text{thin-lens } (yi)}{\text{actual } y}, \qquad c = \frac{u + i}{y}$$

When this is done, we have the following thick-lens paraxial setup:

c	d	n_e
0.56		
	0.30	1.4478604
-0.50963703		
	0.13	1.6163841
0.34125624		
	0.18	1.7044410
0.15858307		
$f'_e = 10.4819,$	$l' = 9.5758$	

Tracing paraxial rays in other wavelengths reveals only very small departures from the thin-lens system. These are caused by the small assumptions that were made in deriving Eq. (37b).

We must next achromatize for the zonal rays by use of the $D - d$ method. For the Δn values, we use $(n_g - n_C)$ because we are endeavoring to unite C, e, and g at a common focus. When this is done, the fourth curvature becomes 0.14697738, and the focal length drops to 9.7209. However, the spherical aberration is found to be $+0.35096$, and we must bend the lens to the right to remove it. Repeating the design with $c_1 = 0.6$, and adding the marginal, zonal, and paraxial rays in all three wavelengths gives the spherochromatism curves shown in Fig. 70a. Both the zonal aberration and the spherochromatism are clearly excessive, and so we adopt the device of introducing a narrow air space after the front element. As this quickly undercorrects the spherical aberration, we return to the above setup, with the addition of an air space, and once more determine the last radius by the $D - d$ method:

c	d	n_e
0.56		
	0.3	1.4478604
-0.50963703		
	0.026	(air)
-0.50963703		
	0.13	1.6163841
0.34125624		
	0.18	1.7044410
0.15895303		
$f'_e = 9.9959,$	$l'_e = 9.0037$	

The spherochromatism curves are shown in Fig. 70b, and the whole situation is greatly improved. This is about as far as we can go. Increasing the air space still further would lead to a considerable overcorrection of the zonal residual, and the result would be worse instead of better.

It is of interest to compare this apochromatic system with a simple doublet made from ordinary glasses. An $f/8$ doublet was therefore designed using the regular procedure, the glasses being

	n_C	n_e	n_g
(a) crown:	1.52036	1.52520	1.53415
(b) flint:	1.61218	1.62115	1.63887

The final system was

c	d	
0.2549982		
	0.2	(crown)
−0.2557933		
	0.1	(flint)
0.00964734		

and the spherochromatism curves are shown in Fig. 70c. It is clear that the zonal aberration is negligible, the only real defect being the secondary spectrum. However, the effort to correct this in the three-lens apochromat has increased the zonal aberration and spherochromatism so much that it is doubtful if the final image would be actually improved thereby. An apochromat is useful only if some means can be found to eliminate the large spherochromatism that is characteristic of such systems.

It should be noted that this solution is far from being the only possible triple apochromat that can be designed. We could assemble the three elements of our thin-lens solution in any order; we could introduce an airgap in the other interface; and of course we could use quite a different set of glasses. Anyone seriously engaged in designing such a system is well advised to try out some of these other possibilities.

CHAPTER 7

Oblique Pencils

An oblique pencil of rays from an extraaxial object point contains meridional rays that can be traced by the ordinary computing procedures already described, and also a large number of skew rays that do not lie in the meridian plane. Each skew ray intersects the meridian plane at the object point and again at a "diapoint" in the image space, and nowhere else. Skew rays require special ray-tracing procedures, which will be discussed in Section III. These are much more complex than for a meridional ray, and skew rays were seldom used before the advent of electronic computers; now they are routinely traced by all lens designers.

I. PASSAGE OF AN OBLIQUE BEAM THROUGH A SPHERICAL SURFACE

A. COMA AND ASTIGMATISM

When a light beam is refracted obliquely through a spherical surface, several new aberrations arise that do not appear on the lens axis. To understand why this is so, we may consider the diagram in Fig. 71, showing a

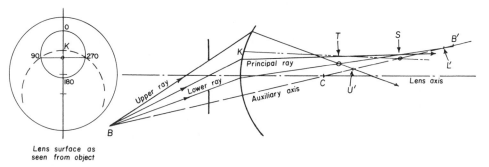

FIG. 71. Origin of coma and astigmatism.

single refracting surface and an aperture stop that admits a circular cone of rays from an off-axis object point B. We label the rays through the rim of the aperture by their position angles taken clockwise from the top as viewed

137

from the image space, so that the upper ray is called 0° and the lower ray 180°, while the rear and front sagittal rays become 90° and 270°, respectively. The line joining the object point B to the center of curvature of the surface, C, is called the "auxiliary axis," and obviously there is complete rotational symmetry around this axis just as there is rotational symmetry around the lens axis for an axial object point. Moreover, because of this symmetry, every ray from the object point B passing through the aperture stop must cross the auxiliary axis somewhere in the image space. If we could trace a paraxial ray from B along the auxiliary axis, it would form an image of B at, say, B'. However, because of the spherical aberration arising at the surface, the intersection point for all other rays will move along the auxiliary axis toward the surface by an amount proportional to the square (approximately) of the height of incidence of the ray above the auxiliary axis. Thus the upper limiting ray might cross the auxiliary axis at, say, U', and the lower limiting ray at L'. It is at once evident that the upper and lower rays do not intersect each other on the principal ray but in general above or below it; the height of the intersection point above or below the principal ray is called the "tangential coma" (a relic of the old custom of calling meridional rays tangential because they form a tangential focal line.)

To find the point at which the two sagittal rays at 90° and 270° intersect the auxiliary axis, we note that these rays are members of a hollow cone of rays centered about the auxiliary axis, all coming to the same focus on that axis. The upper ray of this hollow cone strikes the refracting surface at K, slightly higher than the principal ray, so that the spherical aberration of this ray will be a little greater than that of the principal ray, forming an image at S on the auxiliary axis. Since S lies below the principal ray on our diagram this indicates the presence of some negative sagittal coma, but not as much as the tangential coma that we found previously. Indeed, it can be shown[1] that for a very small aperture and obliquity, the tangential coma is three times the sagittal coma; the exaggerations in our diagram do not make this relation obvious, but at least both comas do have the same sign.

We thus see that the extreme upper and lower rays of the marginal zone come to a focus at T, while the extreme front and rear rays come to a different focus at S. The longitudinal separation between S and T is called the astigmatism of the image, and evidently both coma and astigmatism arise whenever a light beam is refracted obliquely at a surface. It is essential to note that each surface in a lens has a different auxiliary axis, and that the proportion of coma and astigmatism therefore varies from surface to surface. It is thus possible to correct coma and astigmatism independently in a lens system provided there are sufficient degrees of freedom available.

---

[1] A. E. Conrady, pp. 284, 742.

B. VIGNETTING

In many lenses, and particularly those having a considerable axial length, an oblique pencil may be unable to traverse the lens without part of the beam being obstructed by the end lens apertures. For instance, in the Triplet lens shown in Fig. 72 the upper rays of the 20° oblique beam are cut off by

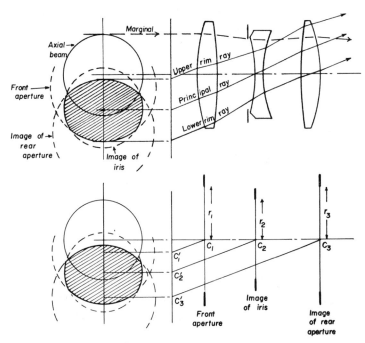

FIG. 72. Vignetting diagrams.

the rear lens aperture, and the lower rays by the front aperture, so that the beam fails to fill the iris. This process is known as "vignetting," the oblique beam projected on to a plane perpendicular to the axis in the object space having the shape shown. Vignetting is one of the reasons why the illumination on the film in a camera falls off at increasing transverse distances from the lens axis. Other reasons are (a) the $\cos^4$ law, (b) distortion of the entrance pupil at high obliquities, and (c) image distortion. The effects of these various factors have been discussed elsewhere.[2]

To plot the vignetting diagram of a lens, the locations of the upper and lower "rim" rays are readily found by trial, but it is then necessary to

---

[2] R. Kingslake, Illumination in optical images, in "Applied Optics and Optical Engineering," Vol. II, p. 195. Academic Press, New York, 1965.

determine the radii of the upper and lower limiting circular arcs. The lower arc obviously has the same radius as the front lens aperture, but the upper arc is the image of the rear aperture as seen through the lens. Its radius bears the same ratio to the radius of the entering axial beam as the diameter of the rear aperture itself bears to the diameter of the axial beam as it emerges from the rear of the lens.

In addition to the circles corresponding to the front and rear lens apertures, an oblique beam is limited also by the iris, and the image of the iris must therefore be projected into the object space along with the image of the rear aperture. To locate this iris image, we add a ray parallel to the upper and lower rim rays and passing through the center of the iris. This middle ray is projected into the vignetting diagram in Fig. 72, and we draw a circle about it having the diameter of the entering axial beam because the axial beam necessarily fills the iris completely. The vignetted area of the oblique beam is shown shaded. The "vignetting factor" is the ratio of the area of the oblique beam to the area of the axial beam, both measured in a plane perpendicular to the lens axis. It is, of course, an assumption that the images of the iris and of the rear lens aperture are circles; indeed, they are much more likely to be arcs of ellipses, but we make very little error by plotting them as circles.

An alternative method of plotting the vignetting diagram is shown in the lower diagram of Fig. 72. We begin by determining the location and size of the images of the rear aperture and of the iris, projected into the object space, by use of paraxial rays traced right-to-left from the centers of those two apertures. The front aperture and the two images are shown at C_1, C_2, and C_3, their computed radii being, respectively, r_1, r_2, and r_3. We can now replace the lens by these three circles, and project their centers at any required obliquity onto a vertical reference plane as shown. Knowing the centers of the circles and their radii, it is a simple matter to draw the vignetting diagram directly. Of course, this procedure cannot be as accurate as the first method, but it is much simpler and generally sufficiently accurate for most purposes. This simple procedure cannot be used for wide-angle or fisheye lenses where the pupil is seriously distorted or tilted.

C. The Principal Ray

The principal ray (sometimes called the chief ray) of an oblique pencil at any obliquity is generally regarded as the ray lying midway between the upper and lower limiting rays passing through the lens at that obliquity. It is highly desirable that the physical iris diaphragm or other limiting aperture be placed at the point where the principal ray crosses the lens axis in order that when the lens is stopped down the principal ray is the last to be cut off.

However, if this is not possible for mechanical reasons and because the crossing point often depends on the obliquity, it is best to favor the rays at the highest obliquity, but there may be mechanical reasons why this is not possible.

D. THE ENTRANCE AND EXIT PUPILS

Once the position of the aperture stop has been established, the paraxial images of the stop in the object and image spaces are known as the entrance and exit pupils, respectively, by analogy with the human eye. These pupils constitute the common bases of entering and emerging oblique beams, and for object points close to the lens axis they remain fixed in position and size. At considerable angles from the axis, however, the pupils become distorted and shifted, and at really extreme angles they may even appear tilted and no longer perpendicular to the lens axis. Indeed, without this tilting of the entrance pupil a fisheye lens covering a full $\pm 90°$ in the object space would not transmit any light at the edge of the field. One should be very careful in discussing the pupils of a lens that only the paraxial images of the aperture stop are intended.

II. TRACING OBLIQUE MERIDIONAL RAYS

For any given object point, or for any given obliquity angle if the object is at infinity, a specific meridional ray must be defined by some convenient ray parameter. This may be the height A at which the ray intersects the tangent plane at the first lens vertex, or it may be the intersection length L of the ray relative to the front lens surface. For a ray proceeding uphill from left to right and entering the lens above the axis, the A will be positive and the L negative.

Whatever ray parameter is chosen, it is necessary to use appropriate "opening equations" to convert the given ray data into the familiar (Q, U) values to trace the ray.

1. A Finite Object

If the object point is defined by its H and d_0 (Fig. 73), then

$$\tan U = (A - H)/d_0 \quad \text{and} \quad Q = A \cos U$$

If the ray is defined by its L value, then

$$\tan U = H/(L - d_0) \quad \text{and} \quad Q = L \sin U$$

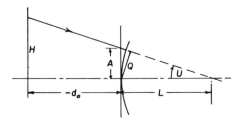

FIG. 73. Opening equations.

2. A Very Distant Object

The slope angle of all entering rays is now the same, being equal to the principal-ray slope U_{pr}; we use only the second of the opening equations to find Q.

3. Closing Equations

Having traced an oblique ray through a lens, we generally wish to know the height at which it crosses the paraxial image plane. This is given by (Fig. 74)

$$H' = (Q' - l' \sin U')/\cos U'$$

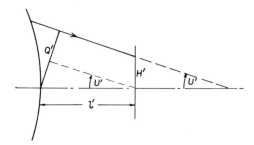

FIG. 74. Closing equations.

4. Intersection of Two Emerging Rays

Sometimes we wish to know the coordinates of the intersection point of two traced rays, knowing their L or their Q' values and also their slope angles U'. The formulas to be used are (Fig. 75a)

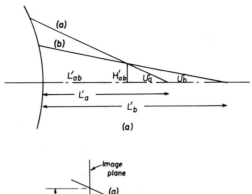

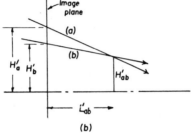

FIG. 75. Coordinates of the intersection of two rays.

$$L'_{ab} = \frac{L'_a \tan U'_a - L'_b \tan U'_b}{\tan U'_a - \tan U'_b}, \qquad \text{where} \quad L' = Q'/\sin U'$$

$$H'_{ab} = (L'_a - L'_{ab}) \tan U'_a = (L'_b - L'_{ab}) \tan U'_b \qquad (61a)$$

A. THE MERIDIONAL RAY PLOT

Having traced a number of oblique rays through a lens from a given object point, we need some way to plot the results and interpret the mixture of aberrations that exists in the image. This mixture will contain spherical aberration, of course, and also the oblique aberrations coma and meridional field curvature. Astigmatism as such will not appear because it involves sagittal rays, which are not traced in a meridional beam. The two chromatic aberrations will not appear unless colored oblique rays are being traced.

The usual procedure is to plot the intercept height H' of the ray at the paraxial image plane as ordinate, with some reasonable ray parameter as abscissa. For the latter we may use the Q value of the ray at the front lens surface, or the incidence height A of the ray at the tangent plane to the front vertex, or the intersection length L of the ray at the first lens surface. Sometimes we use the height of the ray at the paraxial entrance pupil plane, or its height in the stop. However, there is good reason to use as abscissa the tangent of the ray slope angle U' in the image space. When this is done a perfect image point plots as a straight line, whose slope is a measure of the distance from the paraxial image plane to the oblique image point. The reason for this can be seen in Fig. 75b, which shows two rays in an oblique pencil having heights H_a' and H_b' at the image plane and emerging slope angles U_a' and U_b', respectively. The longitudinal distance L_{ab}' from the image plane to the intersection of these rays with one another is given by

$$H_{ab}' = H_a' - L_{ab}' \tan U_a' \quad \text{and} \quad H_{ab}' = H_b' - L_{ab}' \tan U_b'$$

Eliminating H_{ab}' gives

$$L_{ab}' = \frac{H_a' - H_b'}{\tan U_a' - \tan U_b'} \tag{61b}$$

If the data of the two rays are plotted on a graph connecting H' with $\tan U'$, the slope of the line joining the two ray points will be a direct measure of L_{ab}'. Consequently if all the rays in the beam have the same L_{ab}', their ray points will all lie on a straight line, with the lower rim ray at the left and the upper at the right. The principal ray will fall about midway between the two rim rays. A perfect lens with a flat field will plot as a horizontal straight line (Fig. 76a). A perfect lens with an inward-curving field plots as a

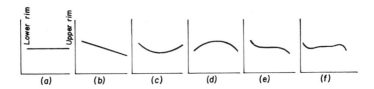

FIG. 76. Some typical $H - \tan U$ curves. (a) A perfect lens; (b) inward-curving field; (c) positive coma; (d) negative coma; (e) spherical undercorrection; (f) zonal spherical aberration.

straight line sloping down from left to right (b). Primary coma is represented by a parabolic graph, the ends being up in the case of positive coma (c) and down for negative coma (d). Primary spherical aberration is represented by a cubic curve, and if the image along the principal ray lies in the paraxial image plane, the middle of the cubic curve will be horizontal (e). Zonal spherical aberration is revealed by a curve with a double bend, which is a combination of a cubic curve for the primary aberration component and a fifth-order curve for the secondary aberration (f). Of course, any imaginable mixture of these aberrations can occur, and the experienced designer soon gets to recognize the presence of the different aberrations by the shape of the curve.

III. TRACING A SKEW RAY

A skew ray is one that starts out from an extraaxial object point and enters a lens in front of or behind the meridian plane. It should be noted that for every skew ray there is another skew ray that is an image of the first, formed as if the meridian plane were a plane mirror. Thus, having traced one skew ray we have really traced two, the ray in front of the meridian plane and the corresponding ray behind it. These two skew rays intersect each other at the same diapoint (see Fig. 6).

In tracing a skew ray, we denote a known point on the ray as $X_0 Y_0 Z_0$, and the direction cosines of the ray as K, L, M. Of course, in the object space the point $X_0 Y_0 Z_0$ can be the original object point, and we must somehow specify the direction cosines of the particular entering ray that we wish to trace. This is often done by specifying the point at which the entering ray pierces the tangent plane at the first lens vertex. Then, knowing $X_0 Y_0 Z_0$ and K, L, M we can determine the point X, Y, Z at which the ray strikes the following lens surface, and after refraction it will have a new set of direction cosines $K'L'M'$ and proceed on its way. The ray-tracing problem thus reduces to two steps: the transfer of the ray from some known point to the next surface, and the refraction of the ray at the next surface.

A. TRANSFER FORMULAS

Since the direction cosines of a line are defined as the differences between the X, Y, Z coordinates of two points lying on the line divided by the distance between these points, it is clear from Fig. 77 that

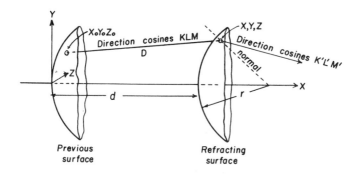

FIG. 77. Transfer of a skew ray from one surface to the next.

$$K = \frac{X - X_0 + d}{D}, \qquad L = \frac{Y - Y_0}{D}, \qquad M = \frac{Z - Z_0}{D}$$

where D is the distance along the ray from the point $X_0\, Y_0\, Z_0$ to the point of incidence X, Y, Z, and d is the axial separation of the surfaces. By means of these relationships we see that

$$X = KD + (X_0 - d), \qquad Y = LD + Y_0, \qquad Z = MD + Z_0 \quad (62a)$$

The equation of the next refracting surface is, of course, known. For a sphere of radius r it is

$$X^2 + Y^2 + Z^2 - 2rX = 0 \qquad (62b)$$

and substituting Eqs. (62a) in (62b) gives the equation to be solved for D as

$$D^2 - 2rF \cdot D + rG = 0$$

where

$$F = K - \frac{K(X_0 - d) + LY_0 + MZ_0}{r}$$

$$G = \frac{(X_0 - d)^2 + Y_0^2 + Z_0^2}{r} - 2(X_0 - d) \qquad (63)$$

The solution is, of course,

$$D = r\left[F \pm \left(F^2 - \frac{G}{r} \right)^{1/2} \right]$$

The ambiguous sign of the root indicates the two possible points of intersection of the ray with a complete sphere of radius r. Only one of these is useful, and the appropriate sign must be chosen. Remember that D must always be positive. Knowing D we return to Eq. (62a) and calculate X, Y, and Z, the coordinates of the point of incidence. For a plane surface,

$$D = G/2F = -(X_0 - d)/K$$

B. THE ANGLES OF INCIDENCE

It is a well-known property of direction cosines that the angle between two intersecting lines is given by

$$\cos I = Kk + Ll + Mm$$

Here K, L, M are the direction cosines of the ray and k, l, m the direction cosines of the normal at the point of incidence. For a spherical surface

$$k = 1 - \frac{X}{r}, \qquad l = -\frac{Y}{r}, \qquad m = -\frac{Z}{r} \qquad (64)$$

Hence

$$\cos I = F - \frac{D}{r}$$

$$\cos I' = [1 - (n/n')^2(1 - \cos^2 I)]^{1/2} \tag{65}$$

For a plane, $\cos I = K$.

C. REFRACTION EQUATIONS

To derive the refraction equations, we refer back to the diagram (Fig. 8) used in connection with the process of graphical ray tracing. It is reproduced in Fig. 78. In the vector triangle OAB, OA is a vector of magnitude n in the

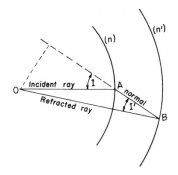

FIG. 78. Refraction of a skew ray.

direction of the incident ray, OB is a vector of magnitude n' in the direction of the refracted ray, while AB is a vector of magnitude $n' \cos I' - n \cos I$ in the direction of the normal. Hence we may construct the vector equation

$$n'\mathbf{R}' = n\mathbf{R} + (n' \cos I' - n \cos I)\mathbf{N}$$

where $\mathbf{R}'$, $\mathbf{R}$, and $\mathbf{N}$ are unit vectors. Since the components of a unit vector are simply the direction cosines of the vector, we can resolve the vector equation into its three component equations:

$$n'K' = nK + (n' \cos I' - n \cos I)k$$

$$n'L = nL + (n' \cos I' - n \cos I)l \tag{66}$$

$$n'M' = nM + (n' \cos I' - n \cos I)m$$

The direction cosines of the normal, k, l, and m, are given in Eq. (64). Hence Eq. (66) becomes

$$n'K' = nK - J(X - r)$$
$$n'L = nL - JY \qquad\qquad (67)$$
$$n'M' = nM - JZ$$

where $J = (n' \cos I' - n \cos I)/r$. As a check on our work, we can verify that $(K'^2 + L^2 + M'^2) = 1$. For refraction at a plane surface these relations become

$$K = \cos I, \qquad K' = \cos I'$$
$$n'L = nL, \qquad n'M' = nM, \qquad J = 0$$

D. TRANSFER TO THE NEXT SURFACE

This has already been described. The direction cosines $K'L'M'$ become the new K, L, M, and we calculate the new point of incidence by Eqs. (63), (62), (65), and (67), in order.

E. OPENING EQUATIONS

1. *Distant Object*

Here we have a parallel beam incident upon the lens inclined at an angle Upr to the lens axis. Then

$$K = \cos Upr, \qquad L = -\sin Upr, \qquad M = 0$$

The point of incidence of the particular skew ray must be determined in some way so that the X, Y, Z can be found. It is common to define the ray by its point of incidence with the tangent plane at the vertex of the first surface. If this is done, it is convenient to regard this tangent plane as the first lens surface with air on both sides of it, and use the general transfer equations to go from the tangent plane to the first refracting surface in the ordinary way.

2. *Near Object*

Here again we assume a tangent plane at the first lens surface, and we specify the point Y, Z at which the skew ray is to pierce that plane. The $X_0 Y_0 Z_0$ of the object point are, of course, known and also the distance d between the object and the front vertex. Then

$$K = \frac{d}{D}, \qquad L = \frac{Y - Y_0}{D}, \qquad M = \frac{Z - Z_0}{D}$$

where

$$D^2 = d^2 + (Y - Y_0)^2 + (Z - Z_0)^2$$

F. Closing Equations

The closing equations for a skew ray are trivial, since the ray can be transferred to the final image plane by the ordinary transfer equations. This process gives the Y', Z' coordinates of the intersection of the ray with the image plane directly. The d is, of course, nothing but the back focal distance from the rear vertex of the lens to the image plane.

G. Diapoint Location

For some purposes we may desire to determine the diapoint location of the skew ray. As has been stated, this is the point where the ray pierces the meridian plane. The Z coordinate of the diapoint is therefore zero, but the other coordinates must be found. By means of a diagram such as that in

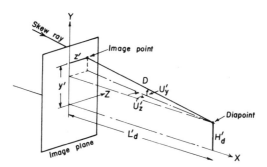

Fig. 79. Diapoint calculation.

Fig. 79 it is easy to show that

$$L'_d = -Z'K'/M' \qquad \text{and} \qquad H'_d = Y' - (Z'L/M')$$

where $K'L'M'$ are the direction cosines of the ray as it emerges from the lens, and $Y'Z'$ are the coordinates of the point where the ray pierces the image plane. The L'_d, H'_d are the required coordinates of the diapoint relative to the midpoint of the image plane and the optical axis of the lens.

H. Example of a Skew-Ray Trace

To illustrate the kind of record required in the manual tracing of a skew ray by these formulas, we will trace a ray through our old familiar cemented doublet objective, entering at an upward slope of 3° through a point at unit distance behind the meridional plane and on the same level as the principal ray. Regarding the tangent plane at the first vertex as a refracting surface, the starting data at that surface are

$$X = 0, \qquad K = \cos(-3°) = 0.9986295$$
$$Y = 0, \qquad L = -\sin(-3°) = 0.0523360$$
$$Z = 1, \qquad M = 0$$

We now transfer the ray from the tangent plane to the first spherical refracting surface in the usual way. The results of the trace are shown in Table X.

TABLE X

MANUAL TRACING OF A SKEW RAY

	Tangent plane				Image plane
r	∞	7.3895	-5.1784	-16.2225	∞
d	0	1.05	0.4	11.28584	
n	1	1.517	1.649	1	
$(n/n')^2$		0.4345390	0.8463106	2.719201	
F	0.9986295	0.8000638	0.9673926	0.9952001	
G	0.1353271	1.584704	0.9077069	22.627015	
D	0.0680706	0.8939223	0.4623396	11.368061	
X	0	0.0679773	-0.0895928	-0.0276675	0
Y	0	0.0035625	0.0342546	0.0491091	0.6289086
Z	1	1.0	0.9584830	0.9457557	-0.0033456
$\cos I$		0.9894178	0.9726889	0.9958924	
$\cos I'$		0.9954155	0.9769360	0.9887907	
J		0.0704550	-0.0261467	0.0402796	
K	0.9986295	0.9983305	0.9991040	0.9952011	
L	0.0523360	0.0343342	0.0321289	0.0510025	
M	0	-0.0464436	-0.0275280	-0.0834884	

IV. GRAPHICAL REPRESENTATION OF SKEW-RAY ABERRATIONS

A. THE SAGITTAL RAY PLOT

The name "sagittal" is generally given to the 90 and 270° skew rays that lie in a plane perpendicular to the meridian plane, containing the principal ray. This is not one single plane throughout a lens but it changes its tilt after each surface refraction. The point of intersection of a sagittal ray with the paraxial image plane may have both a vertical error and a horizontal error relative to the point of intersection of the principal ray, and both these errors can be plotted separately against some suitable ray parameter. This parameter is often the horizontal distance z from the meridian plane to the point where the entering ray pierces the tangent plane at the first lens vertex. Figure 80 shows a typical set of meridional and sagittal ray plots for a triplet

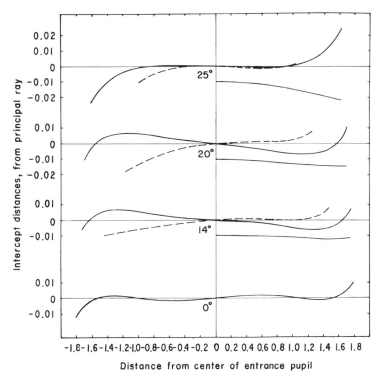

FIG. 80. Typical ray plots for a triplet lens.

photographic objective.[3] The meridional plot, of course, has no symmetry, but the two sagittal ray plots do have symmetry. The vertical errors are identical for rays entering at equal distances in front of and behind the meridian plane, these errors being forms of sagittal coma. The horizontal errors are antisymmetrical, so that the error of the 90° ray is equal and opposite to the horizontal error of the 270° ray; these errors represent

[3] C. Baur and C. Otzen, U.S. Patent 2,966,825, filed Feb. 1957.

sagittal field curvature and sagittal oblique spherical aberration, strictly analogous to the effects of tangential field curvature and tangential oblique spherical aberration in the ordinary meridional ray plot.

B. A Spot Diagram

The meridional and sagittal ray plots already discussed take account of only the rays passing through a cross-shaped aperture over the lens. To include every possible skew ray, it is necessary to divide the lens aperture into a checkerboard of squares and to trace a ray through every intersection of the lines. Assuming that each ray carries the same amount of light energy, the assembly of the intersection points of all such rays with the paraxial focal plane will be a fair representation of the type of image that may be expected when the lens has been made up and tested. Actually it requires a large number of rays, say, over 100, to provide a fair approximation to the actual image. Of course, it is unnecessary to trace rays both behind and in front of the meridional plane since they are identical, but it is necessary to plot both rays in the image plane. Such dot patterns are called spot diagrams, and they were obviously never plotted before the advent of high-speed computers to do the ray tracing. A typical spot diagram for an $f/2.8$ triplet photographic objective is shown in Fig. 81; for this pattern no less than 380 skew rays were traced through each side of the lens aperture.

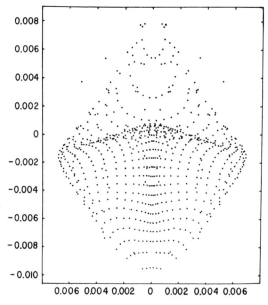

FIG. 81. A typical spot diagram for an $f/2.8$ triplet lens at 14° off axis ($f' = 10$).

C. Encircled Energy Plot

By counting the rays enclosed by a succession of circles of increasing size laid over the spot diagram, it is possible, by a suitable computer program, to plot a graph of the "encircled energy" of a given lens at several obliquities. This assumes that each ray carries the same amount of light energy, a justifiable assumption if the rays are incident in a checkerboard pattern at the entrance pupil of the lens. To include chromatic effects it is necessary to trace many rays in other colors, the size of the checkerboard squares for each wavelength being dependent on the spectral response of the detector intended for that particular lens.

The encircled energy graphs of the $f/2.8$ triplet used above are shown in

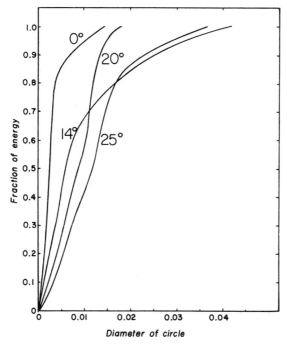

FIG. 82. Encircled energy plot.

Fig. 82. As can be seen, one sheet of graph paper shows the entire performance of the lens at several obliquities. Graphs of this type provide the designer with a great deal of useful information, particularly in comparing one design with another. Some designers prefer MTF graphs, but this type of calculation generally leads to so many graphs that the result is more likely to confuse than to help.

V. RAY DISTRIBUTION FROM A SINGLE ZONE OF A LENS

The nature of the various oblique aberrations of a lens may be better understood if we trace a family of rays passing through a single zone of a lens, both on and off axis. We take the cemented telescope doublet used many times before and isolate a single zone of radius one unit. On axis, all the rays from this zone will, of course, intersect at a single point forming a perfect focus. At an obliquity of only one degree, however, the rays from the zone form a succession of complicated loop figures as shown in Fig. 83. As

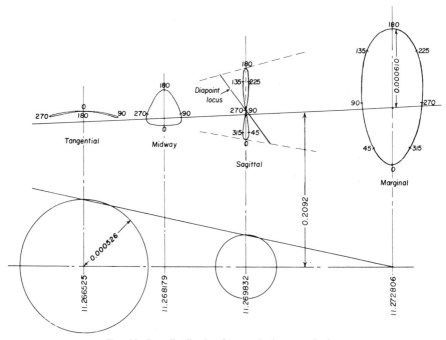

FIG. 83. Ray distribution from a single zone of a lens.

before, the upper and lower rim rays are labeled 0 and 180°, while the sagittal rays are 90 and 270°. Referring to this diagram, the tangential focus is at the intersection of rays 0 and 90°, giving 0.000084 for the amount of tangential coma. The sagittal focus is where the 90 and 270° rays intersect, forming a sagittal coma of magnitude 0.000035, about one-third of the tangential coma. It can be proved that in the absence of higher-order aberrations this ratio should be exactly 3 : 1. The presence of field curvature is indicated by the tangential and sagittal foci not being in the same plane as the axial image of the zone. This series of patterns arises at each zone of the lens, and it is clear that when all zones are open together, the resulting image is very complicated indeed.

It should be noted that the diapoint locus of the zone is also indicated in Fig. 83. It is bisected by the sagittal image, and it ends at the upper and lower tangential rays. Conrady refers to the diapoint locus as the "characteristic focal line" of a zone.

CHAPTER 8

Coma and the Sine Condition

I. THE OPTICAL SINE THEOREM

The optical sine theorem is the equivalent for marginal rays of the Lagrange theorem, which applies only to paraxial rays. The sine theorem provides an expression for the image height formed by a pair of sagittal rays passing through a single zone of a lens. It is valid for a zone of any size but only at very small obliquity. This obliquity limitation effectively removes all aberrations except coma, which is represented by a difference between the image height for the selected zone and the paraxial image height given by the theorem of Lagrange.

To derive the optical sine theorem we consider the perspective diagram in Fig. 84a, which shows a pair of sagittal rays passing through a single

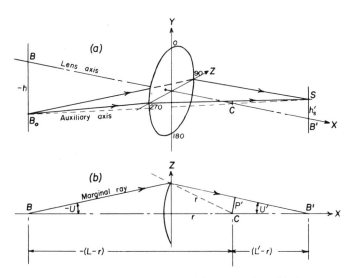

FIG. 84. Derivation of the sine theorem. (a) Oblique view; (b) plan view.

refracting surface, and Fig. 84b, which shows the path of the marginal ray through the same zone. The entering and emerging marginal ray slopes are U and U', respectively, in the usual way.

157

It was pointed out on p. 138 that a pair of sagittal rays intersect on the auxiliary axis drawn through the object point and the center of curvature of the surface. Hence, by the similar triangles shown in Fig. 84,

$$\frac{h'_s}{h} = \frac{CS}{CB_0} = \frac{L' - r}{L - r} = \left(\frac{P'}{\sin U'}\right)\left(\frac{\sin U}{P}\right) \qquad \text{(by Fig. 19a)}$$

$$= \frac{n \sin U}{n' \sin U'}$$

Hence

$$h'_s n' \sin U' = hn \sin U \qquad (68)$$

It is essential to remember that h'_s is the height of the sagittal image for the zone, i.e., the intersection of the 90 and 270° rays, and has no relation whatsoever to the height of any other rays from the zone. It is, in particular, not related to the meridional rays in any way.

II. THE ABBE SINE CONDITION

Abbe regarded coma as a consequence of a difference in image height from one lens zone to another, and he thus realized that a spherically corrected lens (in his case a microscope objective) would be free from coma near the center of the field if the marginal and paraxial magnifications $m = nu/n'u'$ and $M = n \sin U/n' \sin U'$ were equal, that is,

$$u/u' = \sin U/\sin U' \qquad (69)$$

This is known as Abbe's sine condition.

For a very distant object, the sine condition takes a different form. As was shown on p. 52, the Lagrange equation for a distant object is

$$h' = -(n/n') \tan U_{pr} f' \qquad (23)$$

where f' is the distance from the principal plane to the focal point measured along the paraxial ray, or $f' = y_1/u'_k$. A similar expression can be written for the focal length of a marginal ray, namely,

$$F' = Y_1/\sin U'_k \qquad (70)$$

where F' is the distance measured along the marginal ray from the equivalent refracting locus to the point where the ray crosses the lens axis. Thus for a spherically corrected lens and a distant object, Abbe's sine condition reduces to

$$F' = f'$$

This relation tells us that in such a lens, called by Abbe an "aplanat," the equivalent refracting locus is part of a hemisphere centered about the focal point. The maximum possible aperture of an aplanat is therefore $f/0.5$, although this aperture is never achieved in practice. The greatest practical aperture is about $f/0.65$ when the emerging ray slope is about $50°$.

There is no equivalent rule for a lens that is aplanatic for a near object, such as a microscope objective. We can, if we wish, assume that in such a case the two principal planes are really parts of spheres centered about the axial conjugate points, but we could just as easily make any other suitable assumption provided the marginal ray moves from one principal "plane" to the other along a line lying parallel to the lens axis, as indicated for paraxial rays in Fig. 23.

If the refractive index of either the object space or the image space is other than 1.0, we must include the actual refractive index in the f-number:

$$f\text{-number} = \frac{\text{focal length } f'}{\text{entering aperture } 2y}\left(\frac{n}{n'}\right)$$

Thus if the image space were filled with a medium of refractive index 1.5, the highest possible relative aperture would be $f/0.33$. To realize the benefit of this high aperture the receiver, film or photocell, must be actually immersed in the dense medium. Similarly, when a camera is used for underwater photography, the effective aperture of the lens is reduced by a factor of 1.33, which is the refractive index of water.

A. Coma for the Three Cases of Zero Spherical Aberration

It was shown on p. 104 that there are three cases in which a spherical surface has zero spherical aberration: (a) when the object is at the surface itself, (b) when the object is at the center of curvature of the surface, and (c) when the object is at the aplanatic point. It so happens that each of these possible situations also satisfies the Abbe sine condition, thus justifying the name aplanatic for all of them. The reason for this is that in each case the ratio $\sin U/\sin U'$ is a constant. Thus, we have the following:

Case (a), object at surface: $U = I$, $U' = I'$; hence $\sin U/\sin U' = n/n'$
Case (b), object at center: $U = U'$; hence $\sin U/\sin U' = 1$
Case (c), object at aplanatic point: $I = U'$, $I' = U$; hence $\sin U/\sin U' = n'/n$

The aplanatic single-lens elements discussed in Chapter 5, Section I,B, are corrected for both spherical aberration and coma, and hence fully justify the name aplanatic. It should be added that such a lens introduces both chromatic aberration and astigmatism in the sense that would be expected from a single positive element.

III. OFFENSE AGAINST THE SINE CONDITION, OSC

It is clear that we ought to be able to derive some useful information about the magnitude of the coma from a knowledge of the paraxial and marginal magnifications, even though the lens does have some spherical aberration. This situation is indicated in Fig. 85. In this diagram B' repre-

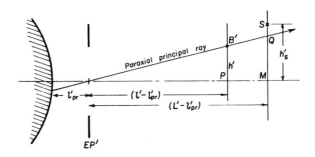

FIG. 85. Offense against the sine condition.

sents an oblique image point in the paraxial image plane of a lens at very small obliquity, its height h' being given by the Lagrange equation. The point S represents the sagittal image formed by a single zone of the lens, its height h'_s being computable by the sine theorem. The point S is assumed to lie in the same focal plane as the marginal image M. At the very small obliquity considered here, the principal ray must be traced by paraxial formulas; it emerges through the center of the exit pupil as shown.

We may express the magnitude of the sagittal coma by the dimensionless ratio QS/QM in the marginal image plane, and we call this ratio the "offense against the sine condition," or OSC. Thus

$$OSC = \frac{QS}{QM} = \frac{SM - QM}{QM} = \frac{SM}{QM} - 1$$

The length SM is the h'_s given by the sine theorem; the length QM is obtainable from the paraxial image-height h' by

$$QM = h'\left(\frac{L - l'_{pr}}{l' - l'_{pr}}\right)$$

Hence

$$OSC = \frac{h'_s}{h'}\left(\frac{l' - l'_{pr}}{L' - l'_{pr}}\right) - 1$$

For a near object we can insert the values of h' and h'_s by the Lagrange and sine theorems, respectively, giving

$$OSC = \frac{u' \sin U}{u \sin U'}\left(\frac{l' - l'_{pr}}{L' - l'_{pr}}\right) - 1$$

$$= \frac{M}{m}\left(\frac{l' - l'_{pr}}{L' - l'_{pr}}\right) - 1 \tag{71}$$

where M and m are, respectively, the image magnification for the finite and paraxial rays.

The bracketed quantity, which contains data relating both to the spherical aberration of the lens and the position of the exit pupil, can be readily modified to

$$\left(1 - \frac{LA'}{L' - l'_{pr}}\right)$$

and for a very distant object, M/m can be replaced by F'/f'. Hence for a distant object, Eq. (71) becomes

$$OSC = \frac{F'}{f'}\left(1 - \frac{LA'}{L' - l'_{pr}}\right) - 1 \tag{72}$$

Conrady[1] states that the maximum permissible tolerance for *OSC* is 0.0025 for telescope and microscope objectives. This large tolerance is because in those instruments the object of principal interest can always be moved into the center of the field for detailed study. A very much smaller tolerance applies to photographic objectives.

A. SOLUTION FOR STOP POSITION FOR A GIVEN *OSC*

Since the exit-pupil position (l'_{pr}) appears in the formulas for *OSC*, it is clear that as we shift the stop along the axis the *OSC* will change provided there is some spherical aberration in the lens. If the spherical aberration has been corrected, then shifting the stop will have no effect on the coma. We can thus solve for the value of l'_{pr} to give any desired *OSC* by inverting Eqs. (71)

---

[1] A. E. Conrady, p. 395.

and (72). For a near object,

$$l'_{pr} = L' - \frac{LA'}{(\Delta m/M) - (m\ OSC/M)}$$

For a distant object,

$$l'_{pr} = L' - \frac{LA'}{\Delta F/F' - (f'\ OSC/F')}$$

These formulas find use in the design of simple eyepieces and landscape lenses for low-cost cameras.

B. SURFACE CONTRIBUTION TO OSC

By a process similar to that used for determining the surface contribution to spherical aberration (p. 101), we can develop a formula giving the surface contribution to OSC. For this derivation, we trace a marginal ray and the paraxial principal ray. Then the development given on p. 101 indicates that in our present case we have

$$(Snu_{pr})'_k - (Snu_{pr})_1 = \sum (Q - Q')ni_{pr} \tag{73}$$

We can see from the diagram in Fig. 86 that $S' = (L' - l'_{pr})\sin U'$, and si-

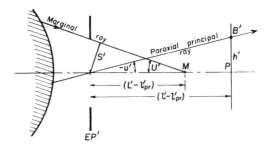

FIG. 86. Surface contribution to OSC.

milarly for the incident ray. Hence, dividing Eq. (73) by the Lagrange invariant and substituting for S and S' we get

$$\left[\frac{(L' - l'_{pr})\sin U'n'u'_{pr}}{h'n'u'}\right]_k - \left[\frac{(L - l_{pr})\sin Unu_{pr}}{hnu}\right]_1 = \sum \frac{(Q - Q')ni_{pr}}{(h'n'u')_k} \tag{74}$$

Now $h'/u'_{pr} = -(l' - l'_{pr})$, and $h/u_{pr} = -(l - l_{pr})$. Also, by the Lagrange and sine theorems we have

$$\left[\frac{\sin U'}{u'}\right]_k = \frac{hn \sin U}{h'_s n'}\left(\frac{h'n'}{hnu}\right) = \frac{\sin U_1}{u_1}\left(\frac{h'}{h'_s}\right)_k$$

Substituting all this in (74) gives

$$-\left[\left(\frac{L-l'_{pr}}{l'-l'_{pr}}\cdot\frac{h'}{h'_s}\right)_k \frac{\sin U_1}{u_1}\right] + \left[\frac{L-l_{pr}}{l-l_{pr}}\cdot\frac{\sin U}{u}\right]_1 = \sum \frac{(Q-Q')ni_{pr}}{(h'n'u')_k} \quad (75)$$

Now by Fig. 85 we see that

$$\left(\frac{L-l'_{pr}}{l'-l'_{pr}}\cdot\frac{h'}{h'_s}\right)_k = \frac{QM}{SM} = \frac{SM-QS}{SM} = 1 - \frac{coma'_s}{SM} = 1 - OSC \quad \text{(approx.)}$$

Thus Eq. (75) becomes

$$(OSC-1) + \left(\frac{L-l_{pr}}{l-l_{pr}}\right)_1 = \frac{u_1}{\sin U_1}\sum \frac{(Q-Q')ni_{pr}}{(h'n'u')_k}$$

and hence

$$OSC = \left[1 - \frac{L-l_{pr}}{l-l_{pr}}\right]_1 + \frac{u_1}{\sin U_1}\sum \frac{(Q-Q')ni_{pr}}{(h'n'u')_k}$$

$$= \frac{-LA_1}{(l-l_{pr})_1} + \frac{u_1}{\sin U_1}\sum \frac{(Q-Q')ni_{pr}}{(h'n'u')_k} \quad (76)$$

It should be noted that any spherical aberration in the object leads to a contribution to the *OSC*. Also, the factor outside the summation, $u_1/\sin U_1$, becomes y_1/Q_1 for a distant object.

Example. As an example in the use of this contribution formula, we will take our old familiar telescope doublet and trace a paraxial principal ray through the front vertex at an entering angle of, say, $-5°$ ($\tan = -0.0874887$), giving

y_{pr}		0	0.0605558	0.0821525
$(nu)_{pr}$	-0.0874887	-0.0874887	-0.0890323	-0.0857457
u_{pr}	-0.0874887	-0.0576721	-0.0539916	
$i_{pr} = (y_{pr}c - u_{pr})$		0.0874887	0.0459782	0.0489275

$$l'_{pr} = -0.9580946$$

$$OSC = \frac{u'}{\sin U'}\left(\frac{l'-l'_{pr}}{L'-l'_{pr}}\right) - 1 = -0.000171$$

For the OSC contribution formula, we pick up the data of the marginal ray from p. 28, giving the following tabulation:

$Q - Q'$	-0.017118	-0.022061	0.037258	
n	1	1.517	1.649	
i_{pr}	0.0874887	0.0459782	0.0489275	
Constant	5.715023	5.715023	5.715023	
OSC contribution	-0.008559	$-.008794$	0.017179	$\Sigma = -0.000173$

For this formula, the Lagrange invariant has the value $(h'n'u') = 0.1749774$. The excellent agreement between the direct and contribution calculations is evident.

C. ORDERS OF COMA

The coma of a pencil of rays at finite aperture and field may be analyzed into orders, as follows:

$$
\begin{aligned}
\text{coma} = \quad & a_1 Y^2 H' + a_2 Y^4 H' + a_3 Y^6 H' + \cdots \\
+ & b_1 Y^2 H'^3 + b_2 Y^4 H'^3 + b_3 Y^6 H'^3 + \cdots \\
+ & c_1 Y^2 H'^5 + c_2 Y^4 H'^5 + c_3 Y^6 H'^5 + \cdots \\
+ & \cdots
\end{aligned}
$$

The first term, $a_1 Y^2 H'$, is the primary term, and it evidently varies as aperture squared and obliquity to the first power. The whole top row of terms is included in the OSC, because OSC is applicable to any aperture but only to a small field. The higher-order terms represent forms of coma that appear in photographic lenses of high aperture at angles of considerable obliquity.

D. THE COMA G-SUM

There is a G-sum expression for the primary coma of a thin lens analogous to that for primary spherical aberration.[2] It varies with aperture squared and image height to the first power. The coma of the object, if any, is transferred to the final image by the ordinary transverse magnification, whereas primary spherical aberration, being a longitudinal quantity, is transferred by the longitudinal magnification rule. It should be noted that this coma G-sum expression is valid only if the stop is located at the thin lens. The formula is

$$
\text{coma}'_s = \text{coma}_s(h'/h) + h'y^2(-\tfrac{1}{4}G_5 cc_1 + G_7 cv_1 + G_8 c^2) \tag{77}
$$

---

[2] A. E. Conrady, p. 324.

where

$$G_5 = 2(n^2 - 1)/n \qquad G_7 = (2n + 1)(n - 1)/2n = G_2/n$$
$$G_8 = n(n - 1)/2 = G_1/n$$

As before, with a thin doublet we assume an infinitely thin air layer between the elements, and then the G-sums may be directly added. Hence

$$OSC = \text{coma}'_s/h' = y^2[(G\text{-sum})_a + (G\text{-sum})_b]$$

E. Spherical Aberration and OSC

It should be clear by now that the spherical aberration of a lens is determined by the location of the intersection point of a ray with the lens axis, whereas the coma is determined by the slope angle of the ray at the image. If the shape of a lens is such that the equivalent refracting locus is too flat, the marginal focal length will be too long and the OSC will be positive. A thin lens bent to the left meets this condition. Similarly, if the rim of the lens is bent to the right the OSC will be negative. Plotting spherical aberration and OSC against bending for such a lens gives a graph like the one in Fig. 87.

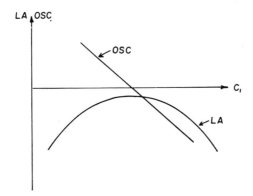

Fig. 87. Typical effect of bending a single thin lens.

It should be noted that in any reasonably thin lens, the lens bending for which the spherical aberration reaches an algebraic maximum is almost exactly the same bending as that which makes the OSC zero. For the primary aberrations of a single thin lens, this is easily verified by comparing the value of c_1 that makes $\partial LA'_p/\partial c_1 = 0$ [Eq. (55)] with the value of c_1 that makes the $\text{coma}_p = 0$ [Eq. (77)]. It will be found that for a variety of refractive indices and a variety of object distances, the c_1 for zero coma is always slightly greater than the c_1 for maximum spherical aberration.

There is, of course, no aperture limit for a nonaplanatic system. A parabolic mirror, for example, has zero spherical aberration for a distant axial object point, but the focal length of each ray is the distance from the mirror surface to the image point, measured along the ray. The focal length continuously increases with increasing incidence height, and the image is afflicted with enormous positive coma.

CHAPTER 9

Design of Aplanatic Objectives

It has already been mentioned that Abbe used the term aplanatic to refer to a lens system corrected for both spherical aberration and *OSC*. We shall use the term "aplanat" for a relatively thin spherically corrected achromat that is also corrected for *OSC* and thus satisfies the Abbe sine condition. As we have seen, a cemented doublet has three degrees of freedom, which are used to maintain the focal length and control the spherical and chromatic aberrations. To include *OSC* correction requires an additional degree of freedom, which can be obtained in various ways. The principal types of aplanat will now be considered.

I. BROKEN-CONTACT TYPE

In this type of aplanat the powers of the two lens elements are determined for chromatic correction by the ordinary (c_a, c_b) formulas, and then each element is separately bent to correct the spherical aberration and *OSC*. Obviously such a lens cannot be cemented, and this type is used mainly in large sizes. It is possible to perform a thin-lens predesign by the use of Seidel aberration contributions, but the subsequent insertion of finite thicknesses causes such an upset that the preliminary study turns out to be useless.

Since bending a lens affects the spherical aberration most, the *OSC* much less, and the chromatic aberration scarcely at all, we select the bending of an achromatic doublet that corresponds to the peak of the spherical-aberration curve, since this is known to be close to the zero-*OSC* form. We then make small trial bendings of each element separately, and plot a double graph by which we can correct *LA'* and *OSC* using the bending parameters c_1 and c_3.

Following this procedure, we see that the graph in Fig. 65 for a crown-in-front achromat reaches its maximum at $c_1 = 0.15$. Then, since $c_a = 0.5090$, we find $c_2 = c_1 - c_a = -0.3590$. We start with $c_3 = c_2$ and a narrow air space such as 0.01. Suitable lens thicknesses are 0.42 and 0.15 for an aperture of 2.0 and a trim diameter of 2.2 (f/5). We shall, of course, achromatize every trial system by solving for the last radius by the $D - d$ method. Our starting system A is found to have

$$LA' = 0.1057, \qquad OSC = 0.00062$$

the l'_{pr} for the *OSC* formula being taken as zero, i.e., the stop is assumed to be in contact with the rear surface.

To build up our double graph, we next apply trial bendings of 0.01 to each of the two lens elements separately. Bending the crown element gives B, with

$$LA' = 0.1057, \qquad OSC = -0.00270$$

Restoring c_1 to its initial value and bending the flint element gives C, with

$$LA' = -0.1245, \qquad OSC = 0.00304$$

These values are plotted in Fig. 88, with *LA'* as abscissa and *OSC* as ordin-

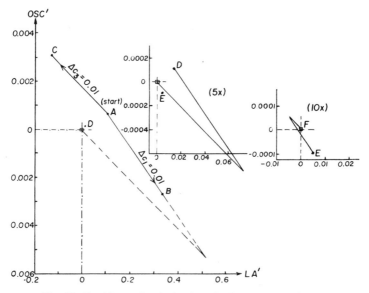

FIG. 88. Double-graph solution for a broken-contact aplanat.

ate. Inspection of the graph suggests that we ought to reach the aim point (0, 0), assuming that all aberrations are linear, by applying to the original setup:

$$\Delta c_1 = 0.0172, \qquad \Delta c_3 = 0.0212$$

These changes, with $c_a = 0.5090$ and the usual solution of the last radius, give system D, with

$$LA' = 0.01522, \qquad OSC = 0.00010$$

The coma is satisfactory but the spherical aberration is still much too large.

Since we are too close to the aim point for the first graph to be useful, we enlarge both scales by a factor of 5, and drawing lines parallel to the original lines suggests that we try

$$\Delta c_1 = 0.0025, \qquad \Delta c_3 = 0.0031$$

Applying these changes gives system E, with

$$LA' = 0.0052, \qquad OSC = -0.00010$$

To remove these residuals we draw an even larger-scale graph (10 ×), giving

$$\Delta c_1 = -0.00045, \qquad \Delta c_3 = -0.00020$$

The final system F has $LA' = -0.00027$ and $OSC = 0$. The zonal aberration is $+0.0040$, of the unusual overcorrected type. As was pointed out on p. 132, this is to be expected in view of the narrow air space in this lens.

The final system is (Fig. 89) given in the following tabulation:

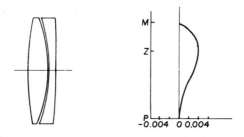

FIG. 89. Broken-contact aplanat.

c	d	n_D	V
0.16925			
	0.42	1.523	58.6
−0.33975			
	0.01	(air)	
−0.33490			
	0.15	1.617	36.6
−0.06682			

where $f' = 9.9302$, $l' = 9.6058$, $Y = 1.0$, trim $= 2.2$, $LA' = -0.00027$, $LZA = +0.0040$, $OSC = 0$.

It is worth noting that the air space in this lens has the form of a negative element, the equivalent of a positive glass lens that undercorrects the spherical aberration. A broken-contact lens of this type requires the utmost care in mounting, and particularly in centering one element relative to the other. In a large lens it is best to mount each element into a separate metal ring, using push–pull screws to secure and adjust the separation to give the best possible definition. For a small lens, the air space is too narrow for a loose spacer to be used, and it is best to mount the two elements on opposite sides of a fixed metal flange with separate clamping rings to hold them in place.

II. PARALLEL AIR-SPACE TYPE

As an alternative to the broken-contact type just discussed, we may prefer to keep the two inner radii equal to save the cost of a pair of test plates, and vary the air space to correct the spherical aberration. Then if the coma is excessive, we can correct it by bending the whole lens.

As before, we start at the maximum of the bending curve, with $c_1 = 0.15$ and $c_a = 0.5090$, giving $c_2 = c_3 = -0.3590$. In Section I our starting setup had an air space of 0.01, giving $LA' = 0.10566$ and $OSC = 0.00062$ (setup A in Fig. 88). If we increase the air space to 0.04, with the usual $D - d$ solution for the last radius, we obtain setup B:

$$LA' = -0.01466, \qquad OSC = 0.00305$$

We next apply a trial bending of 0.01 to the entire lens, with the 0.04 air space, and we get $LA' = -0.00646$ and $OSC = 0.00201$ (setup C). These values are plotted on a double graph with spherical aberration as abscissa and OSC as ordinate, as before (Fig. 90). Evidently a further bending by 0.0198 should bring us close to the aim point. Actually this bending gave $LA' = 0.00014$ and $OSC = 0$ (setup D).

As the zonal aberration of this air-spaced lens is liable to be strongly overcorrected, we prefer to have a small negative value for the marginal aberration. Since our trial change in air space gave $\partial LA'/\partial(\text{space}) = 4.0$, we try increasing the air space by 0.0001. This gives the final setup as follows for trim diam. $= 2.2$, $f' = 10.1324$, $l' = 9.7012$, $LA' = -0.00017$, $LZA = +0.00666$, $OSC = 0$. It should be noted that the overcorrected zonal aberration is now 1.6 times as great as in the broken-contact design, and this is the principal reason why the previous type is generally to be preferred. However, the air space is now wider, which may be of help in designing the lens mount.

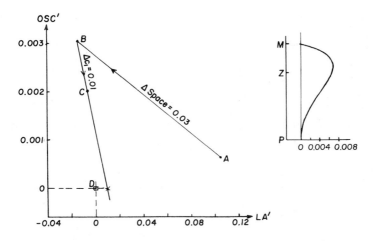

FIG. 90. Double graph for a parallel-air space aplanat.

c	d	n	V
0.1798			
	0.42	1.523	58.6
−0.3292			
	0.0401		
−0.3292			
	0.15	1.617	36.6
−0.0553			

III. AN APLANATIC CEMENTED DOUBLET

On p. 165 it was pointed out that the bending of a cemented doublet that yields zero OSC almost coincides with the bending for maximum spherical aberration. Consequently, if we can find two types of glass for which the spherical aberration curve just reaches zero at the top of the bending parabola, this peak bending will also be very nearly aplanatic.

Some guidance as to likely types of glass can be obtained by calculating the spherical G sums, and plotting the thin-lens bending curve as in Fig. 65, relying on the fact that the true thick-lens curve coincides closely with the approximate thin-lens graph shown there. A few trials along these lines indicates that the spherical aberration curve will be bodily lowered if we increase the V difference between the glasses or if we reduce the n difference between them.

Since we have only three degrees of freedom in a cemented doublet, which must be used for focal length, spherical aberration, and OSC, it is clear that we must leave the choice of glass until the end in order to secure achromatism by the $D - d$ method. Since there are more crowns than flints in the Schott catalog, we will adopt some specific flint and try several crowns to see how the chromatic condition is operating. Taking as our flint Schott's SF-19 with $n_D = 1.66662$ and $V = 33.08$, we select three possible crowns, and with each we adopt an approximate value of $c_a = 0.3755$ for $f' = 10$. Thus we find by a series of trials the value of c_3 that corrects the spherical aberration at $f/5$. The whole lens is then bent, again by a series of trials, to eliminate the OSC. Then the $D - d$ sum of the aplanat is found for the marginal ray; finally we find, also by the $D - d$ method, what value the crown V should have to produce a perfect achromat. Repeating the process with each of the three crowns enables us to plot a locus of possible crowns on the glass chart (Fig. 41), and if this locus happens to pass through an actual glass, that glass will be used to complete the design.

Our three trials give the results (Fig. 91) in the following tabulation:

Crown glass type	n_D	$n_F - n_C$	V_D	$\sum (D - d) \, \Delta n$	Desired crown V for perfect achromatism
SK-12	1.58305	0.00983	59.31	0.0000365	58.16
BaK-6	1.57436	0.01018	56.42	−0.0000906	59.23
SK-11	1.56376	0.00928	60.75	0.0000083	60.46

Even without plotting a curved locus on the glass chart, we can see that the third selection, SK-11, gives a close achromat with our chosen flint. The final design after solving the last radius to give a zero $D - d$ sum is as follows:

c	d	Glass	n_D	V_D
0.1509				
	0.32	SK-11	1.56376	60.75
−0.2246				
	0.15	SF-19	1.66662	33.08
−0.052351				

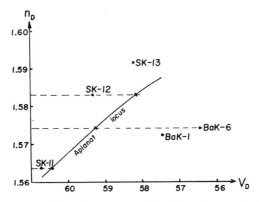

FIG. 91. Locus of crown glasses for a cemented doublet aplanat, using SF-19 as a flint.

for trim = 2.2, $f' = 10.3663$, $l' = 10.1227$, $LA' = -0.00010$, $LZA = -0.00176$, $OSC = 0.00060$. This would make an excellent objective. The undercorrected zonal residual is very small, being only 40% of that found for the common achromat on p. 127. The smallness of this zone is due to the use of higher-index glasses. It is found that with a given flint, the zonal residual is large for low-index crowns, drops to a minimum for some medium-index crowns, and then rises again for high-index crowns. There are clearly two opposing tendencies here. Raising the crown index weakens the front radius, but it also lowers the index difference across the cemented interface, requiring a stronger curvature there.

IV. A TRIPLE CEMENTED APLANAT

Another way to obtain the additional degree of freedom necessary for OSC correction is to divide the flint component of a cemented doublet into two and place one part in front of and the other part behind the crown component to make a cemented triplet. Of course, alternatively, the crown component could be divided in this way, but it is generally better to divide the flint.

Conrady[1] has given a very complete study of this system on the basis of the spherical and coma G sums. To apply such an analysis, for each element we need net curvature c, bending parameter c_1, and reciprocal object distance v. In this triplet lens we have two absolute degrees of freedom, which we

---

[1] A. E. Conrady, p. 557.

may call x and y: the amount of flint power in the front element, and a bending of the whole lens. We therefore define $x = c_1 - c_2 = c_a$, and $y = c_2$.

The total powers of crown and flint are found by the ordinary (c_a, c_b) formulas on p. 79; they will be referred to here as Cr and Fl. Hence for the three thin-lens elements we have the following:

	Lens a	Lens b	Lens c
Net curvature (c)	x	Cr	Fl $- x$
Front curvature (c_1)	$y + x$	y	$y -$ Cr
Reciprocal object distance (v)	$v_1 = 1/l_1$	$v_1 + (n_a - 1)x$	$v_1 + (n_a - 1)x + (n_b - 1)$Cr

Here $n_a = n_c$ is the flint index and n_b the crown index. In order to draw a section of the lens to determine suitable thicknesses, we note that the thin-lens value of c_4 is $x + y - (\text{Cr} + \text{Fl})$.

In performing the G sum analysis, we find that the spherical aberration expression is a quadratic, while the coma expression is linear. Hence there will be two solutions to the problem. To reduce the zonal residual and to have as many lenses as possible on a block, we choose that solution in which the strongest surface has the longer radius.

As an example, we will design a low-power cemented triplet microscope objective, with magnification 5 × and tube length 160 mm. This represents a focal length of 26.67 mm. The numerical aperture, $\sin U'_4$, is to be 0.125; therefore the entering ray slope is $\sin U_1 = -0.025$. We will use the following common glass types:

(a) Flint: F-3, $n_e = 1.61685$, $\Delta n = 0.01659$, $V_e = 37.18$
(b) Crown: BaK-2, $n_e = 1.54211$, $\Delta n = 0.00905$, $V_e = 59.90$

with $V_b - V_a = 22.72$. The (c_a, c_b) formulas give for the total crown and flint powers

$$\text{Cr} = 0.1824, \qquad \text{Fl} = -0.0995$$

Conrady's G-sum analysis gives the following approximate solutions:

x:	-0.088	-0.019
y:	$+0.158$	$+0.072$

Hence		
$c_1 = x + y$:	0.070	0.053
$c_2 = y$:	0.158	0.072
$c_3 = y - \text{Cr}$:	-0.0244	-0.1104
$c_4 = x + y - (\text{Cr} + \text{Fl})$:	-0.0129	-0.0299 (or -0.03035 by $D - d$)

The strongest curve in the first solution is $c_2 = 0.158$, whereas the strongest surface in the second solution is $c_3 = -0.1104$. We therefore continue work on the second solution. Since the radii are approximately 18.9, 13.9, -9.1, and -33.4, we can draw a diagram of the lens. The semiaperture is to be 5.0 since the Y of the marginal ray is $160 \times 0.025 = 4.0$. Suitable thicknesses are found to be 1.0, 3.5, and 1.0, respectively (Fig. 92), all dimensions in millimeters.

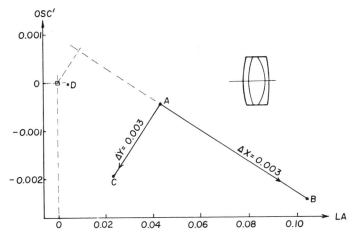

FIG. 92. Double graph for a cemented triple aplanat.

We begin by tracing a marginal ray with $L_1 = -160$ and $\sin U_1 = -0.025$, solving the last radius by the $D - d$ method as usual. We calculate the LA' and OSC of this ray, (setup A) assuming that the aperture stop is located at the rear lens surface, whence $OSC = (Ml'/mL) - 1$. We then make small experimental changes in x and y, and plot the usual double graph with OSC as ordinate and LA' as abscissa. This graph indicates the required very small changes in x and y from the starting setup to make both aberrations zero, namely, $\Delta x = -0.0017$ and $\Delta y = +0.0014$. These changes give the following solution to the problem:

	c	d	n_e
	0.0527		
		1.0	1.61685
	0.0734		
		3.5	1.54211
	-0.1090		
		1.0	1.61685
$(D - d)$	-0.030667		

with $L_1 = -160$ mm, $\sin U_1 = -0.025$, $LA' = 0.00042$ mm, $LZA = -0.04688$ mm, $OSC = -0.00002$, $l' = 30.145$ mm, $1/m = -4.927$,

$$NA = 0.123.$$

Since the numerical aperture is slightly below our desired value of 0.125, we may shorten the focal length in the ratio $0.123/0.125 = 0.984$ by strengthening all the radii in this proportion. The small zonal residual is far less than the Rayleigh tolerance of 0.21 mm and is negligible.

V. APLANAT WITH A BURIED ACHROMATIZING SURFACE

The idea of a "buried" achromatizing surface was suggested by Paul Rudolph in the late 1890s.[2] Such a surface has glass of the same refractive index on both sides, but because the dispersive powers are different, it can be used to control the chromatic aberration of the lens. Thus achromatism can be left to the end and the available degrees of freedom can be used for the correction of other aberrations.

Some possible matched pairs of glasses from the Schott catalog are as follows:

		n_D	$n_F - n_C$	V_D	V difference
(1)	SK-16	1.62031	0.01029	60.28	
	F-9	1.62030	0.01634	37.96	22.32
(2)	SK-5	1.58905	0.00962	61.23	
	LF-2	1.58908	0.01438	40.97	20.26
(3)	SSK-2	1.62218	0.01171	53.13	
	F-13	1.62222	0.01725	36.07	17.06
(4)	SK-7	1.60720	0.01021	59.47	
	BaF-5	1.60718	0.01233	49.24	10.23

An exact index match is unnecessary, particularly since the actual index of purchased glass may depart from the catalog values by as much as 0.0001.

As an example in the use of a buried surface, we will design a triple aplanat in which the third surface will be buried. The remaining three radii will be used for spherical aberration and coma, the last radius being in all cases solved for the required focal length. We will maintain a focal length of 10.0 and an aperture of $Y = 1.0$ ($f/5$), allowing sufficient thicknesses for a

---

[2] P. Rudolph, U.S. Patent 576, 896, filed July 1896.

trim diameter of 2.2 for the crown and for the insertion of the buried surface in the flint. For the crown lens we will use K-5 glass ($n_D = 1.5224$, $\Delta n = 0.00876$, and $V_D = 59.63$). For the flint we will use the glasses in selection (3) above, performing the ray tracing with the average index of 1.6222. A convenient starting system is:

c	d	n_D
0.16		
	0.42	1.5224
-0.26		
	0.35	1.6222
(solve for f')		

The last curvature comes out to be -0.069605, giving $LA' = 0.01013$ and $OSC = 0$. We now make a trial change in c_1 by 0.01, giving $LA' = 0.02059$ and $OSC = -0.00096$. Returning to the original setup and changing c_2 by 0.01 gives $LA' = -0.00241$ and $OSC = 0.00016$. These values are plotted on the double graph of Fig. 93, and we conclude that we should make a

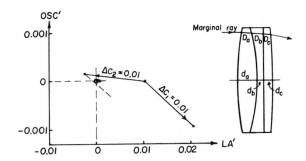

FIG. 93. Double graph for a buried-surface triple aplanat.

further change in c_1 by 0.0022. This completes the design so far as spherical aberration and coma are concerned.

We must now introduce the buried surface for achromatism. To do this we calculate the D values of the two elements along the marginal ray and divide the axial thickness of the second lens suitably, say at 0.15 and 0.20. We tabulate the four surfaces with as much information as we have. Knowing the value of $(D - d) \Delta n$ for the front element, we see that the sum for the remaining two elements must be equal and opposite to it. We also know the sum of the two D values for the last two lenses. Solving the two simultaneous equations tells us that D_b must be 0.2779594 and D_c must be 0.1650274.

Knowing the Y values at the various surfaces, we finally ascertain that c_3 must be 0.0080508. This completes the design, which is as follows:

c	d	n_D	V_D
0.1622			
	0.42	1.5224	59.63
−0.25			
	0.15	1.62222	36.07
0.008051			
	0.20	1.62218	53.13
−0.066105			

for $f' = 10.0$, $l' = 9.6360$, $LA' = -0.00008$, $LZA = -0.00232$, $OSC = -0.00005$.

VI. THE MATCHING PRINCIPLE

If we wish to design an aplanatic lens of such a high aperture that a single doublet is impossible, we resort to the use of two achromats in succession. We now have four degrees of freedom. The subdivision of power between the two components and the air space between them is arbitrary, while the two bendings can be used for the correction of spherical aberration and OSC. Any reasonable types of glass can be used, and by achromatizing each component separately we automatically correct both the chromatic aberration and the lateral color.

The design of lenses of this type has been described in detail by Conrady,[3] in particular when used as a microscope objective of medium power. Having decided upon suitable values for the two arbitrary quantities, we trace a marginal ray through the system from front to back, solving r_3 and r_6 by the $D - d$ method, and we then add two paraxial rays, one through the front component from left to right using $l_1 = L_1$ and $u_1 = \sin U_1$, and the other through the rear component from right to left, taking $u'_6 = \sin U'_6$ and $l'_6 = L'_6$ as starting data. If we can now find such a pair of bendings that the two paraxial rays match in the air space between the lenses, the system will be corrected for both spherical aberration and OSC. This is what is meant by the matching principle.

To make the required trial bendings, we have no problem with the front component, but we must adopt standard entry data for the rear component. We can easily adopt a fixed value for L_4 by always choosing a suitable air

---

[3] A. E. Conrady, p. 662.

space between the lenses, but any standard value of U_4 that we may adopt will never agree exactly with the emerging slope U'_3 from the front component. Consequently it becomes necessary to match actual aberrations in the air space rather than trying to match lengths and angles. So far as lengths are concerned, we have always $L_4 = L'_3 - d$, and we require that $l_4 = l'_3 - d$. Subtracting these tells us that we must select bendings such that

$$LA_4 = LA'_3$$

To match the slope angles of the paraxial rays in the air space, we have approximately that $\sin U_4 = \sin U'_3$, and we require that u_4 should also be equal to u'_3. Dividing these gives

$$OSC_4 = OSC'_3$$

where OSC is defined as

$$OSC = (u/\sin U) - 1$$

Since this kind of OSC does not contain the usual correcting factor for spherical aberration and exit-pupil position, we refer to it as "uncorrected OSC" in the present context.

As an example to illustrate the matching principle, we will design a 10 × microscope objective of numerical aperture 0.25, so that the entering ray slope at the long conjugate end is -0.025. Assuming an object distance of -170 mm, we can trace any desired rays into the front component of the system. It should be noted that, as always, we calculate a microscope objective from the long conjugate to the short, because the long conjugate distance is fixed while the short is not, so that the long-conjugate end becomes the "front" of our system. This conflicts with ordinary microscope parlance, which regards the front of a microscope objective as the short-conjugate end; this is a unique exception and we shall ignore it here.

Our first problem is to deal with the two arbitrary degrees of freedom, namely, the subdivision of refracting power between the two components, and the air space between them. For this, it is common to require that the paraxial ray suffer equal deviation at each component, and to place the rear component approximately midway between the front component and its image. This makes the object distance for all rear-element bendings about 20 mm, and we shall adopt that value here. As the overall paraxial deviation is $0.25 + 0.025 = 0.275$, we must allow each component to deviate the paraxial ray by 0.1375, which makes the ray slope between the components equal to 0.1125. We shall therefore adopt this value of $\sin U_4$ in making all trial bendings of the rear component. For both lenses we use the following common types of glass:

(a) Crown: $n_e = 1.52520$, $n_F - n_C = 0.00893$, $V_e = 58.81$
(b) Flint: $n_e = 1.62115$, $n_F - n_C = 0.01686$, $V_e = 36.84$

with $V_a - V_b = 21.97$. The thin-lens data of the two components are (Fig. 94)

FIG. 94. A Lister-type microscope objective design.

as follows:

Object distance (mm)	Image distance (mm)	Focal length (mm)	Clear aperture (mm)	c_a	c_b	Suitable thicknesses (mm)
−170	37.77	30.90	8.5	0.1649	−0.0874	3.2, 1.0
20.0	9.00	16.36	4.5	0.3116	−0.1650	2.0, 0.8

After determining the last radius by the $D - d$ method in every case, the results of several bendings of each component are found to be as follows:

Front component

$$L_1 = l_1 = -170.0, \qquad \sin U_1 = u_1 = -0.025$$

c_1:	0	0.02	0.04	0.06	0.08
c_3 by $D - d$:	−0.08273	−0.06254	−0.04130	−0.01870	+0.00558
L_3':	33.149	34.666	35.465	35.567	35.005
LA_3':	−0.1474	0.9529	1.3282	0.8596	−0.6049
uncorrected OSC_3':	0.01164	0.03753	0.03727	0.01360	−0.03567

Rear component

$$L_4 = 20.00, \qquad \sin U_4 = 0.1125$$

c_4:	0.05	0.10	0.15	0.20	0.25
c_6 by $D - d$:	−0.11360	−0.05405	0.01450	0.09511	0.19249
$L_6' = l_6'$:	7.3552	7.3700	7.2695	7.0888	6.8570
$\sin U_6' = u_6'$:	0.25862	0.24760	0.23939	0.23259	0.22588
l_4:	18.8706	20.2829	20.7971	20.4545	19.3870
LA_4:	1.1294	−0.2829	−0.7971	−0.4545	0.6130
uncorrected OSC_4:	0.03222	−0.03055	−0.04323	−0.01363	0.05653

These results are plotted side by side on one graph, Fig. 95. It is our aim to

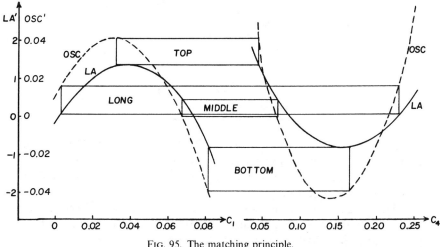

FIG. 95. The matching principle.

select such values of c_1 and c_4 that $LA'_3 = LA_4$ and simultaneously $OSC'_3 = OSC_4$. This is done by searching for rectangles that just fit into the four curves, with spherical aberration and coma points on the same levels. There are four such rectangles to be found, indicating that the curves represent quadratic expressions. The four solutions are as follows:

Rectangle	c_1	c_2	c_3	c_4	c_5	c_6
A (top)	0.032	−0.133	−0.050	0.045	−0.266	−0.119
B (middle)	0.067	−0.098	−0.010	0.070	−0.242	−0.091
C (bottom)	0.081	−0.084	0.007	0.165	−0.147	0.037
D (long)	0.003	−0.162	−0.080	0.228	−0.084	0.147

For many reasons we shall continue the design using solution C. All the other solutions contain stronger surfaces, and moreover both components of solution C contain almost equiconvex crown elements. This starting setup is as follows:

c	d	n_e
0.081		
	3.2	1.52520
−0.08394		
	1.0	1.62115

Cont'd

Cont'd

c	d	n_e
0.00685		
	14.9603	(air)
0.165		
	2.0	1.52520
−0.14654		
	0.8	1.62115
0.03730		

with $l_6' = 7.2095$, $LA_6' = 0.01383$, $u_6' = 0.2361$, $OSC_6' = -0.00297$. For the final OSC_6' calculation we assumed that the exit pupil is in such a position that $(l' - l_{pr}')$ is about 17.0. This puts the exit pupil about 10 mm inside the rear vertex of the objective.

Although this solution is close, we must improve both aberrations by means of a double graph. Changing c_1 by 0.001, and maintaining the $D - d$ solutions and $L_4 = 20.0$, we find that the aberrations become

$$LA_6' = -0.000829, \qquad OSC_6' = -0.002404$$

Restoring the original c_1 and changing c_4 by 0.001 gives

$$LA_6' = 0.001306, \qquad OSC_6' = -0.003279$$

Inspection of the graph suggests that we change the original c_1 by 0.001 and c_4 by −0.01. These changes give

$$LA_6' = -0.000403, \qquad OSC_6' = +0.000474$$

Unfortunately the numerical aperture of the system is now 0.2381, whereas it should be 0.25. We therefore scale all radii down by 4%, which gives

$$LA_6' = 0.001114, \qquad OSC_6' = 0.000221$$

Further reference to the double graph suggests that we try $\Delta c_1 = 0.0005$ and $\Delta c_4 = 0.002$. This change gives the almost perfect solution drawn to scale in Fig. 94, namely,

c	d	n_e
0.08578		
	3.2	1.52520
−0.08576		
	1.0	1.62115
0.009152		
	13.8043	(air)

Cont'd

Cont'd

c	d	n_e
0.16320		
	2.0	1.52520
−0.16080		
	0.8	1.62115
0.02602		

with $l_6' = 6.8925$, $u_6' = 0.2500$, $LA_6' = 0.000004$, $OSC_6' = -0.000095$, $LZA_6' = -0.00289$. In practice, of course, we should apply trifling further bendings to both components to render the crown elements exactly equicon-vex. These changes are so slight as to have no significant effect on any of the aberrations.

The zonal aberration tolerance is $6\lambda/\sin^2 U_m' = 0.053$, so that the zonal residual of our objective is about half the Rayleigh limit. To improve it, we would have to go to a flint of somewhat higher index, but the present design would be acceptable as it stands.

CHAPTER 10

The Oblique Aberrations

In Chapter 7 we introduced the subject of the oblique aberrations of a lens, and discussed in detail the origin and computation of coma. In this chapter we continue the discussion, giving computing procedures for the remaining oblique aberrations, astigmatism, field curvature, distortion, and lateral color.

I. ASTIGMATISM AND THE CODDINGTON EQUATIONS

When a narrow beam of light is obliquely incident on a refracting surface, astigmatism is introduced, and the image of a point source formed by a small lens aperture becomes a pair of focal lines, a series of beam sections being indicated in Fig. 96. One focal line (sagittal) is radial to the

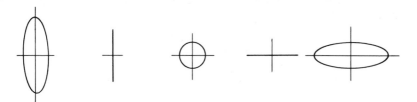

FIG. 96. The astigmatic focal lines.

field and points toward the lens axis, while the other focal line (tangential) is tangential to the field. Both focal lines are perpendicular to the principal ray, and their locations can be calculated once the principal ray has been traced. The astigmatic images formed by the first surface become the objects for the second, and so on through the system. The locations of the focal lines are found by the two Coddington[1] equations, which will now be derived.

A. THE TANGENTIAL IMAGE

In Fig. 97, *BP* is an entering principal ray, *B* being the tangential object

---

[1] H. Coddington, "A Treatise on the Reflexion and Refraction of Light," p. 66. Simpkin and Marshall, London, 1829.

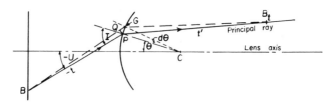

FIG. 97. The tangential focus.

point distant t from the point of incidence P, the length t being measured along the principal ray, negative if the object point lies to the left of the surface as usual. The line BG represents a neighbor ray close to the principal ray, lying in the meridian plane, so close in fact that the short arc $PG = r\, d\theta$ can be regarded as tangent to the refracting surface itself.

The central angle $\theta = U + I$, and hence

$$d\theta = dU + dI \tag{78}$$

The short line PQ, perpendicular to the incident ray, is given by

$$PQ = t\, dU = PG \cos I = r \cos I\, d\theta$$

But by (78) we have $dU = d\theta - dI$. Therefore $PQ = t(d\theta - dI) = r \cos I\, d\theta$, whence

$$dI = \left\{1 - \frac{r \cos I}{t}\right\} d\theta \tag{78a}$$

Similarly for the refracted ray we have

$$dI' = \left\{1 - \frac{r \cos I'}{t'}\right\} d\theta \tag{78b}$$

By differentiating the law of refraction for paraxial rays we obtain

$$n \cos I\, dI = n' \cos I'\, dI' \tag{78c}$$

and inserting (78a) and (78b) into (78c) we get

$$\frac{n' \cos^2 I'_{\mathrm{pr}}}{t'} - \frac{n \cos^2 I_{\mathrm{pr}}}{t} = \frac{n' \cos I'_{\mathrm{pr}} - n \cos I_{\mathrm{pr}}}{r} \tag{79}$$

The term on the right degenerates to the surface power $(n' - n)/r$ when the object point lies on the lens axis so that $I'_{\mathrm{pr}} = I_{\mathrm{pr}} = 0$. It may be regarded as the "oblique power" of the refracting surface for the principal ray. The oblique power is always slightly greater than the axial power, which provides a convenient check on the calculation.

B. THE SAGITTAL IMAGE

The other focal line is located at the sagittal image point B_s. This is a paraxial-type image formed by a pair of sagittal (skew) rays lying close to the principal ray. As explained on p. 138, the image of a point formed by a pair of sagittal rays always lies on the auxiliary axis joining the object point to the center of curvature of the surface. This property of sagittal rays enables us to derive the second Coddington equation locating the sagittal focal line.

In Fig. 98 we show the principal ray, the sagittal object point B, and the

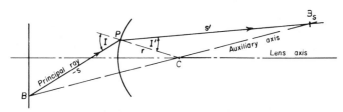

FIG. 98. The sagittal focus.

sagittal image B_s, with the auxiliary axis joining B, C, and B_s. Now, the area of any triangle ABC is given by $\frac{1}{2}ab \sin C$; and since triangle $BPB_s =$ triangle BPC plus triangle PCB_s, we have

$$-\tfrac{1}{2}ss' \sin(180° - I + I') = -\tfrac{1}{2}sr \sin(180° - I) + \tfrac{1}{2}s'r \sin I'$$

whence

$$-ss' \sin(I - I') = -sr \sin I + s'r \sin I'$$

Expanding $\sin(I - I')$ and multiplying by $(n'/ss'r)$ gives

$$-\frac{n' \sin I \cos I' - n' \cos I \sin I'}{r} = -\frac{n' \sin I}{s'} + \frac{n' \sin I'}{s}$$

But by the law of refraction, $n' \sin I'$ can be everywhere replaced by $n \sin I$. When this is done, the $\sin I$ cancels out, giving

$$\frac{n'}{s'} - \frac{n}{s} = \frac{n' \cos I'_{pr} - n \cos I_{pr}}{r} \tag{80}$$

The term on the right is the same oblique power of the surface that we found for the tangential image in Eq. (79). Thus the only difference between the formulas for tangential and sagittal foci is the presence of the $\cos^2$ terms in the tangential formula.

Conrady[2] has also given a very direct derivation of these formulas by a

[2] A. E. Conrady, p. 588.

method depending on the equality of optical paths at a focus. However, the purely geometrical derivations given here are easier to follow and are quite valid.

C. Astigmatic Calculation

To use these formulas to calculate the astigmatism of a lens, we begin by tracing a principal ray at the required obliquity. We calculate the starting values of s and t from the object to the point of incidence measured along the principal ray (see Section C,5).

1. Oblique Power

We next determine the oblique power of each surface by

$$\phi = c(n' \cos I' - n \cos I)$$

for a spherical surface of curvature c. If the surface is aspheric, then it is necessary to calculate the separate sagittal and tangential surface curvatures at the point of incidence by

$$c_s = \sin(U + I)/Y, \qquad c_t = (d^2X/dY^2)\cos^3(U + I)$$

The second derivative d^2X/dY^2 is found from the equation of the aspheric surface. The rest of the data refer to the principal ray itself at the surface. It is common to find a great difference between the sagittal and tangential surface curvatures; indeed, they may even have opposite sign.

2. Oblique Separations

The third step is to calculate the oblique separation between successive pairs of surfaces, measured along the principal ray, by

$$D = (d + X_2 - X_1)/\cos U'_1$$

where the X values of the principal ray at the various surfaces are found by

$$X = \frac{1 - \cos(U + I)}{c}$$

or better by

$$X = G \sin(U + I)$$

3. Sagittal Ray

We then trace the sagittal neighbor ray by applying at each surface

$$s' = \frac{n'}{(n/s) + \phi}$$

Transfer is $s_2 = s'_1 - D$.

4. Tangential Ray

The formula for tracing the tangential neighbor ray is

$$t' = \frac{n' \cos^2 I'}{[(n \cos^2 I)/t] + \phi}$$

Transfer is $t_2 = t'_1 - D$. The process for tracing the tangential ray can be made similar to that for the sagittal ray by listing across the page the values of $n \cos^2 I$ and $n' \cos^2 I'$, and then treating these products as if they were the actual refractive indices of the glasses.

5. Opening Equations

If the object is at infinity, the opening values of both s and t are infinity. If the object is at a distance B from the front lens vertex (negative if to the left), then we must calculate (see Fig. 99a)

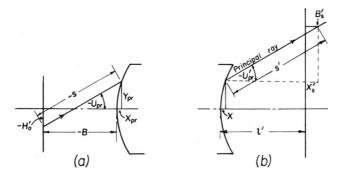

FIG. 99. (a) Opening equations. (b) Closing equations.

$$s = t = (B - X_{pr})/\cos U_{pr} = (Y_{pr} - H_0)/\sin U_{pr}$$

6. Closing Equations

Having traced the sagittal and tangential neighbor rays, we generally wish to know the axial distances of the sagittal and tangential focal lines from the paraxial image plane. These are given by (see Fig. 99b):

$$X'_s = s' \cos U'_{pr} + X - l'$$

$$X'_t = t' \cos U'_{pr} + X - l' \tag{81}$$

where X is the sag of the rear lens surface computed for the principal ray.

Example. As an example in the use of these formulas, we will trace a principal ray through the cemented doublet used several times before, on p. 28. The principal ray will enter the front vertex at an angle of $-3°$. The tabular layout of the computation, as performed on a small pocket calculator, is given in Table XI. The closing equations give $X'_s = -0.02674$ and $X'_t = -0.05641$. The tangential focal line is thus about twice as far from the paraxial focal plane as the radial focal line, and both are inside the focal plane.

TABLE XI

CALCULATION OF ASTIGMATISM ALONG A PRINCIPAL RAY

c		0.1353271	-0.1931098	-0.0616427	
d		1.05	0.4		
n		1.517	1.649		
		Tracing of $-3°$ principal ray			
Q		0	0.0362246	0.0491463	
Q'		0	0.0362270	0.0491086	
I		3.00000	1.57608	1.67722	
I'		1.97708	1.44989	2.76642	
U	-3.0	-1.97708	-1.85089	-2.94009	
		Tabulation of cosines			
$\cos I$		0.9986295	0.9996217	0.9995716	
$\cos I'$		0.9994047	0.9996798	0.9988346	
$\cos U$		0.9994047	0.9994783		0.9986837
		Oblique powers of surfaces			
ϕ		0.0700274	-0.0254994	0.0400344	
		Oblique separations			
X		0	-0.0001268	-0.0000745	
D		1.050499	0.4002611		
		Sagittal ray			
s		∞	20.612450	33.884695	
s'		21.662949	34.284956	11.274028	
		Tangential ray			
$n \cos^2 I$		0.9972609	1.5158524	1.6475874	
$n' \cos^2 I'$		1.5151944	1.6479443	0.9976705	
t		∞	20.586666	33.836812	
t'		21.637165	34.237073	11.244328	

D. GRAPHICAL DETERMINATION OF THE ASTIGMATIC IMAGES

The location of the sagittal focus along a traced principal ray is easy because the image lies on an auxiliary axis drawn from the object point through the center of curvature of the surface.

T. Smith[3] credits Thomas Young with the discovery of a similar procedure for locating the tangential image point. Young's method for the construction of the refracted ray itself involves drawing two auxiliary circles about the center of curvature of the refracting surface, one with radius rn/n' and the other with radius rn'/n (Fig. 100). The incident ray is extended to

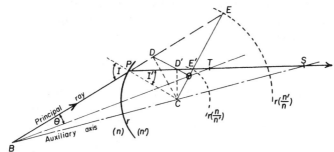

FIG. 100. Young's construction for the sagittal and tangential foci ($n = 1, n' = 1.7$).

cross the second of these auxiliary circles at E, and then E is joined to the center of curvature C. This line crosses the first auxiliary circle at E'; then the refracted ray is drawn from P through E'.

To locate the tangential image of any point B situated on the incident ray, we drop perpendiculars from C onto the two sections of the ray, striking them at D and D', respectively. Then the point of intersection of DD' and EE' is the point O. The line DD' is found to be perpendicular to EE'. This point O replaces C when graphically locating the tangential image of B, so that the line from B to O crosses the refracted ray at the tangential image, while the line from B through the center of curvature C locates the sagittal image.

The proof of this is difficult. It is best to assume that the angle θ in triangle BDO is equal to the angle θ in triangle BPT; then the geometry of the two triangles leads to the regular Coddington equation for the tangential image.

E. ASTIGMATISM FOR THE THREE CASES OF ZERO SPHERICAL ABERRATION

On p. 104 it was pointed out that a single spherical surface contributes no spherical aberration when the object is at (a) the surface itself, (b) the center of curvature of the surface, and (c) at the aplanatic point. On p. 159 it was shown that the OSC also is zero for these three object points.

---

[3] T. Smith, The contributions of Thomas Young to geometrical optics, *Proc. Phys. Soc.* **62B**, 619 (1949).

By means of the Coddington equations it is easy to show that at small obliquity the astigmatism contribution will be zero in cases (a) and (c), but when the object is at the center of curvature the astigmatism contribution is large and in the unexpected sense, i.e., the convex front surface of a positive lens, for instance, contributes positive astigmatism when we would ordinarily have expected it to lead to an inward-curving field. This result is often of great significance, and it explains many anomalies, such as the flat tangential field of a Huygenian eyepiece.

F. ASTIGMATISM AT A TILTED SURFACE

If a lens surface is tilted through a given angle, the procedure outlined on p. 32 can be used to trace the principal ray, and the ordinary Coddington equations can be used to locate the astigmatic images along the principal ray. However, because of the asymmetry, the astigmatism at some angle, say 15°, above the axis will not be the same as the astigmatism at 15° below the axis, and to plot the fields it is now necessary to trace several principal rays with both positive and negative entering obliquity angles.

As an example, we will refer ahead to the design of a Protar lens, p. 274, and pick up the principal-ray data at several obliquities. We will next suppose that the rear lens surface has been tilted clockwise through an angle of 0.10° (6 min of arc), so that $\alpha = 0.1$. By comparing the field curves given in Fig. 101 before and after the last surface was tilted, the effect of the tilt can be readily seen. Briefly, it causes the field to tilt in a counterclockwise direction, the tangential field being tilted and distorted much more than the sagittal field. Limiting ourselves to one field angle, say 17.2°, we find that the tangential field has been tilted by 35.2 min while the sagittal field has been tilted through 13.3 min, both considerably more than the surface tilt that caused the problem. Actually, the effects of a tilt as small as 5 min of arc can generally be detected, and it is customary to try to limit accidental surface tilts in any good lens to about 1 min of arc. Surface tilt does more damage to an image than any other manufacturing error, and in assembling a lens it is essential to avoid tilted surfaces at any cost.

II. THE PETZVAL THEOREM

From very simple considerations, it is clear that a positive lens ought to have an inward-curving field. The extraaxial points on a flat object are further from the lens than the axial point, and consequently their images should be closer to the lens than the axial image, leading at once to an inward-curving field.

The exact amount of this natural field curvature can be calculated by the

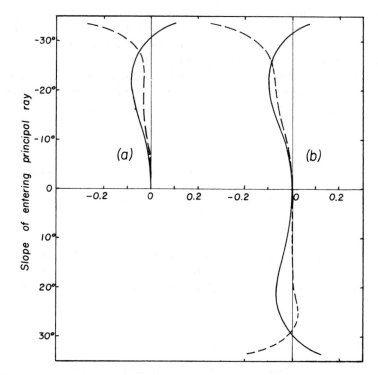

FIG. 101. Fields of a Protar lens. (a) Centered. (b) Rear surface tilted clockwise by 6 min of arc.

following argument. Suppose we place a small stop at the center of curvature C of a single spherical refracting surface (Fig. 102). This will automatically

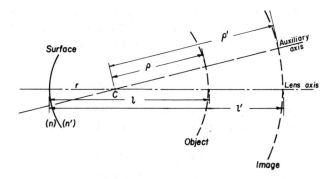

FIG. 102. The Petzval theorem.

eliminate coma and astigmatism by forcing the oblique light to be refracted along an auxiliary axis as if it were an axial beam. If the stop is small enough to eliminate spherical aberration also, we shall be left with nothing but the basic field curvature that we are trying to evaluate.

It is, of course, obvious that under these conditions an object having the form of a sphere centered about C must be imaged as a sphere also centered about C. If the radii of curvature of object and image are represented by ρ and ρ', then

$$\rho = l - r, \qquad \rho' = l' - r$$

and since for a single surface $n'/l' - n/l = (n' - n)/r$, we can readily show that, for one surface,

$$\frac{1}{n'\rho'} - \frac{1}{n\rho} = \frac{n' - n}{nn'r}$$

We can now write this expression for every surface in the lens and add up, but this procedure will be valid only if we can assume that all traces of astigmatism have somehow been eliminated. For such a lens having k surfaces, we find that

$$\frac{1}{n'_k\rho'_k} - \frac{1}{n_1\rho_1} = \sum \frac{n' - n}{nn'r}$$

This expression relates the radius of curvature of the image with the radius of curvature of the object, provided there is no astigmatism present. It is clear, then, that the radius of curvature of the image of a plane object with $\rho_1 = \infty$, is given by

$$\frac{1}{\rho'_k} = n'_k \sum \frac{n' - n}{nn'r} \tag{82}$$

It should be noted that a positive value of ρ corresponds to a negative sag, or an inward-curving image. Hence the sag of the curved image of a plane object, in the absence of astigmatism, will be given by

$$X'_{\text{ptz}} = -\frac{1}{2} h'^2_k n'_k \sum \frac{n' - n}{nn'r} \tag{83}$$

This is the famous Petzval theorem, and we shall have many occasions to refer to it since it is only possible to design a flat-field lens free from astigmatism by reducing the Petzval sum; thus the Petzval theorem dominates the entire design processes for flat-field photographic lenses.

The quantity under the summation in these different expressions is called the Petzval sum, and the radius of curvature of the image is evidently the reciprocal of the Petzval sum. Another useful term is the Petzval ratio, which is the ratio of the Petzval radius to the focal length of the lens. It is given by

$$\rho'/f' = 1/f'\Sigma$$

where Σ is the Petzval sum. Note the reciprocal relationship here. A long-focus lens tends to have a small Petzval sum, while the sum is large in a strong lens of short focal length.

A. Relation between the Petzval Sum and Astigmatism

It can be shown[4] that at very small obliquity angles the tangential astigmatism, i.e., the longitudinal distance from the Petzval surface to the tangential focal line, is three times as great as the corresponding sagittal astigmatism. Thus, if the astigmatism in any lens can be made zero, the two focal lines will coalesce on the Petzval surface. In all other cases the locus of the tangential foci at various obliquities is called the tangential field of a lens, and similarly for the sagittal field. As the Petzval surface in most simple lenses is inward-curving, it is often possible to flatten the tangential field by the deliberate introduction of overcorrected astigmatism, leaving the sagittal image to fall between the Petzval surface and the tangential image. However, when designing an "anastigmat" having a flat field free from astigmatism, it is necessary to reduce the Petzval sum drastically.

If it is necessary to design a lens having an inward-curving field to meet some customer requirement, the astigmatism can easily be removed and the Petzval sum adjusted to give the desired field curvature. On the other hand, if the field must be backward-curving, it is difficult to avoid an excessive amount of overcorrected astigmatism. It is worth noting that in some types of lens, if the Petzval sum is made too small the separation between the astigmatic fields becomes excessively large at intermediate field angles.

B. Methods for Reducing the Petzval Sum

There are three principal methods by which the Petzval sum can be reduced, and one or more of these appear in every type of photographic objective.

1. A Thick Meniscus

If we have a single lens in which both radii of curvature are equal and of the same sign, the Petzval sum will be zero, while the lens power is proportional to the thickness. Cemented interfaces in such a lens have very little

---

[4] A. E. Conrady, p. 739.

effect on the Petzval sum. This property has been used in many symmetrical lenses such as the Dagor and Orthostigmat.

2. Separated Thin Elements

In a system containing several widely separated thin elements, the Petzval sum is given by

$$\text{Ptz} = \sum \phi/n \tag{84}$$

where ϕ is the power of an element. If there is about as much negative as positive power in such a system, the Petzval sum can be made as small as desired. This property has been used in many lenses of the dialyte type (see p. 236).

3. A Field Flattener

An interesting special case is that in which a negative lens element is placed at or near an image plane. This element has little or no effect on the focal length or the aberrations, but it contributes its full power to the Petzval sum (see p. 208). Conversely, if it is necessary to insert a positive lens in an image plane to act as a field lens, then this lens has a large adverse effect on the Petzval sum. For this reason it is almost impossible to reduce the Petzval sum in a long periscope having several internal images. However, by using photographic-type lenses as field lenses it is sometimes possible to reduce the sum appreciably.

It should be noted that in a lens having a long central air space, the Petzval sum is increased if both components are positive (as in the Petzval portrait lens) because the rear component acts partly as a positive field lens. On the other hand, if the rear component is negative (as in a telephoto), then the Petzval sum is reduced, and in an extreme telephoto it may actually becomes negative, requiring some degree of undercorrected astigmatism to offset it.

4. A New-Achromat Combination

The third method for controlling the Petzval sum is to use a crown glass of low dispersion and high refractive index in combination with a flint glass of higher dispersion and a low refractive index. This is precisely opposite to the choice of glasses used in telescope doublets and other ordinary achromats. Lenses of this type are therefore known as "new achromats." They have been used in the Protar (p. 268) and many other types of photographic objectives.

III. DISTORTION

Distortion is a peculiar aberration in that it does not cause any loss of definition but merely a radial displacement of an image point toward or away from the lens axis. Distortion is calculated by determining the height H'_{pr} at which the principal ray intersects the image plane, and comparing that height with the ideal Lagrangian or Gaussian image height calculated by paraxial formulas. Thus

$$\text{distortion} = H'_{pr} - h'$$

where h' for a distant object is given by ($f \tan U_{pr}$), or for a near object by (Hm), where m is the image magnification.

Distortion can be resolved into a series of powers of H', namely,

$$\text{distortion} = aH'^3 + bH'^5 + cH'^7 + \cdots$$

However, very few lenses exhibit much distortion beyond the first cubic term. Because of the cubic law, distortion increases rapidly once it begins to appear, and this makes the corners of the image of a square, for example, stretch out for positive (pincushion) distortion, or pull in with negative (barrel) distortion.

The magnitude of distortion is generally expressed as a percentage of the image height, at the corners of a picture. Figure 103 shows two typical cases

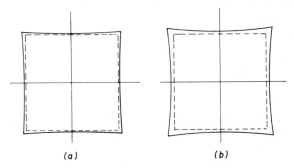

(a)   (b)

FIG. 103. Pincushion distortion. (a) 4%, $d = 0.55$ mm, $r = 565$ mm; (b) 10%, $d = 1.38$ mm, $r = 226$ mm.

of moderate amounts of pincushion distortion, namely, 4 and 10%, respectively. The diagrams represent images that should be 50 mm squares, the quantity d beneath each figure being the lateral displacement of the midpoints of the sides of the square due to distortion. The quantity r is the radius of curvature of the sides of the images, which should, of course, be straight. As can be seen, 4% distortion is just noticeable, whereas 10% is definitely objectionable. Consequently, we generally set the distortion tolerance at about 1% since few observers can detect such a small amount. For specialized applications such as aerial surveying and map copying, the slightest trace of distortion is objectionable, and the greatest care must be taken in the design and manufacture of lenses for these purposes to eliminate distortion completely.

A. MEASURING DISTORTION

Since distortion varies across the field of a lens, it is difficult to determine the ideal Gaussian image height with which the observed image height is to be compared. One method is to photograph the images of a row of distant objects located at known angles from the lens axis and measure the image heights on the film. Since focal length is equal to the ratio of the image height to the tangent of the subtense angle, we can plot focal length against object position and extrapolate to zero object subtense to determine "the" axial focal length with which all the other focal lengths are to be compated. If the lens is to be used with a near object, we substitute object size for angular subtense and magnification for focal length. The determination should be performed at several object positions and the coefficients $a, b, c, \ldots$ found.

B. DISTORTION CONTRIBUTION FORMULAS

To develop an expression for the contribution of each lens surface to the distortion, we repeat the spherical-aberration contribution development on p. 102 but using the principal ray instead of the marginal ray. Thus Eq. (50) becomes

$$(S'n'u')_k - (Snu)_1 = \sum ni(Q - Q')$$

where capital letters now refer to the data of the traced principal ray. Figure 104a shows that at the final image $S'_{pr} = H'_{pr} \cos U'_{pr}$, and similarly for the object. Hence if there are k surfaces in the lens,

$$H' = H\left\{\frac{nu \cos U}{n'_k u'_k \cos U'_k}\right\} + \sum \frac{ni(Q - Q')}{n'_k u'_k \cos U'_k}$$

For a distant object, the first term in this expression reduces to

$$-f'(\sin U/\cos U'_k)_{pr}$$

To relate this formula to the distortion, we note that $Dist = H' - h'$, where h', the Lagrangian image height, is equal to $-f' \tan U_1$. Hence

$$\text{distortion} = h'_{\text{Lagr.}}\left(\frac{\cos U_1}{\cos U'_k} - 1\right)_{pr} + \sum \frac{ni(Q - Q')_{pr}}{n'_k u'_k \cos U'_{k, pr}} \qquad (85)$$

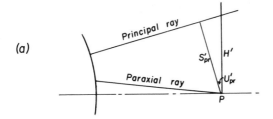

(a)

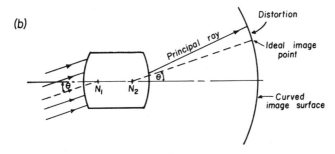

(b)

Fig. 104. Distortion diagrams.

Note that the two parts of this formula are similar in magnitude, the first being caused by the difference in slope of the principal ray as it enters and leaves the system, the second being derived from the lens surface contributions.

To verify the accuracy of this formula, we take the much-used cemented doublet of p. 28 and trace a principal ray entering at $-8°$ through the anterior focal point to form an almost perfectly telecentric system (see Table XII). The agreement in the results of this calculation between the direct measure of the image height and the sum of the various contributions is excellent. For the distortion itself we first calculate

$$h'_{\text{Lagr.}} \left(\frac{\cos U_1}{\cos U'_k} - 1 \right) = -0.0163489$$

When this is added to the sum value we find the distortion to be -0.0618321, again in excellent agreement. The change in slope of the principal ray has contributed about one-third of the distortion, the remainder coming from the lens surfaces themselves.

Unfortunately, the quantities under the summation sign are not really "contributions" that have merely to be added together to give the distortion. Each lens surface, to be sure, provides an amount to be summed, but it

TABLE XII

CALCULATION OF DISTORTION CONTRIBUTIONS

c	0.1353271	-0.1931098	-0.0616427	
d		1.05	0.4	
n		1.517	1.649	
		Paraxial		
ϕ	0.0699641	-0.0254905	0.0400061	
$-d/n$		-0.6921556	-0.2425713	$l' = 11.285857$
y	1	0.9515740	0.9404865	$f' = 12.00002$
nu	0	0.0699641	0.0457080	0.0833332
u	0	0.0461200	0.0277186	0.0833332
$(yc - u) = i$	0.1353271	-0.2298783	-0.0856927	
		$8°$ *Principal ray, with* $L_1 = -11.76$		
Q	1.6527600	1.7050560	1.7263990	
Q'	1.6947263	1.7117212	1.7233187	
U $-8°$		$-0.56367°$	$-2.10291°$	$0.50119°$
		Distortion contributions		
$(Q - Q')_{pr}$	-0.0419663	-0.0066652	0.0030803	
ni	0.1353271	-0.3487254	-0.1413073	
$1/u'_k \cos U'_k$	12.000478	12.000478	12.000478	
Prod.	-0.0681528	0.0278930	-0.0052234	$\sum = -0.0454832$

Hence $H' = 1.6701438 - 0.0454832 = 1.6246606$

also contributes to the slope of the emergent ray in the first term of the distortion expression. The relation just given is therefore mainly of theoretical interest.

C. DISTORTION WHEN THE IMAGE SURFACE IS CURVED

If a lens is designed to form its image on a curved surface, the meaning of distortion must be clearly defined. As always, distortion is the radial distance from the ideal image point to the crossing point of the principal ray; but now the ideal image is represented by the point of intersection of a line drawn through the second nodal point at the same slope as that of a corresponding ray entering through the first nodal point (Fig. 104b). Then

$$\text{distortion} = [(Y_2 - Y_1)^2 + (X_2 - X_1)^2]^{1/2}$$

where subscript 2 refers to the traced principal ray and subscript 1 to the ideal ray through the nodal points.

IV. LATERAL COLOR

Lateral color is similar to distortion in that it is calculated by finding the height of intercept of principal rays at the image plane, but now we must compare two principal rays in two different wavelengths, typically the C and F lines of hydrogen, although, of course, any other specified lines can be used if desired. Then

$$\text{lateral color} = H'_F - H'_C$$

Lateral color can be resolved into a power series, but now there is a first-order term that does not appear in distortion:

$$\text{lateral color} = aH' + bH'^3 + cH'^5 + \cdots$$

Some people consider that only the first term represents lateral color, all the others being merely the chromatic variation of distortion. No matter how it is regarded, lateral color causes a radial chromatic blurring at image points away from the lens axis. Of course, both distortion and lateral color vanish at the center of the field.

A. PRIMARY LATERAL COLOR

The first term of this series, representing the primary lateral color, can be calculated by a method similar to the calculation of OSC, except that now we trace paraxial rays in C and F light instead of tracing a marginal and a paraxial ray in brightest light. Thus, writing paraxial data in F in place of the original marginal ray data, and paraxial data in C in place of the original paraxial ray data, our formula (71) becomes

$$CDM = \frac{\text{lateral color}}{\text{image height}} = \frac{u'_C}{u'_F}\left(\frac{l'_C - l'_{pr}}{l'_F\, l'_{pr}}\right) - 1 \qquad \text{for a near object}$$

$$= \frac{\Delta f'}{f'} - \frac{\Delta l'}{l' - l'_{pr}} \qquad \text{for a distant object} \qquad (86)$$

where $\Delta f' = f'_F - f'_C$ and $\Delta l' = l'_F - l'_C$. The latter is, of course, the ordinary paraxial longitudinal chromatic aberration. The expression CDM is an abbreviation for chromatic difference of magnification and it is strictly analogous to OSC.

In a symmetrical lens, or any other lens in which the pupils coincide with the principal planes, $l' - l'_{pr} = f'$, and the second part of (86) becomes

$$CDM = (\Delta f' - \Delta l')/f' \qquad (87)$$

The numerator of this expression is simply the distance between the second principal planes in C and F light. Thus, if these principal planes coincide, there will be no primary lateral color. This is often a convenient computing device for use in the early stages of a design. Later, of course, it is necessary to trace true principal rays in F and C and calculate the difference in the heights of these rays at the focal plane.

The logic of this last relationship can be understood by the diagram in

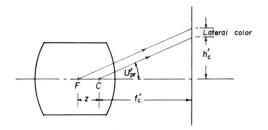

FIG. 105. Primary lateral color depends on z.

Fig. 105, which shows the principal rays in C and F, at small obliquity, emerging from their respective principal points and proceeding to the image plane. It is clear that

$$\text{primary lateral color} = z \tan U'_{pr} = z(h'/f')$$

and hence

$$CDM = \text{lateral color}/h' = z/f'$$

B. APPLICATION OF THE $(D - d)$ METHOD TO AN OBLIQUE PENCIL

It has been shown by Feder[5] that Conrady's $D - d$ method can be applied to an oblique pencil through a lens. He pointed out that if we calculate $\sum D \, \Delta n$ along each ray of the pencil and $\sum d \, \Delta n$ along the principal ray, then we can plot a graph connecting $\sum (D - d) \, \Delta n$ as ordinate against $\sin U'$ of the ray as abscissa. The interpretation of this graph is that the ordinates represent the longitudinal chromatic aberration of each zone, while the slope of the curve represents the lateral color of that zone.

Typical curves at 0 and 20°, calculated for the $f/2.8$ triplet used in plotting the spot diagram shown in Fig. 81, are given in Fig. 106 for $\Delta n = (n_F - n_C)$. The fact that the axial graph is not a straight line indicates the presence of spherochromatism; this is shown plotted in the ordinary way in Fig. 107. The tilt of the 20° curve at the principal-ray point (Fig. 106) indicates the presence of lateral color, of amount about -0.0018. The lateral color found

[5] D. P. Feder, Conrady's chromatic condition, *J. Res. Nat. Bur. Std.* **52**, 47 (1954); res. paper 2471.

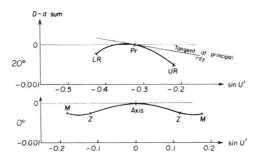

FIG. 106. Application of the $(D - d)$ method to an axial and an oblique pencil through a triplet objective.

by actual ray tracing was $H'_F - H'_C = -0.00168$, which is in excellent agreement considering the difficulty in determining the exact tangent to the curve at the principal-ray point.

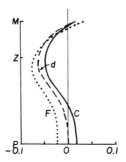

FIG. 107. Spherochromatism of $f/2.8$ triplet objective.

V. THE SYMMETRICAL PRINCIPLE

A fully symmetrical (holosymmetrical) system is one in which each half of the system, including the object and image planes, is identical with the other half, so that if the front half is rotated through 180° about the center of the stop it will coincide exactly with the rear half.

Such a fully symmetrical system has several interesting and valuable properties, notably complete absence of distortion and lateral color, and absence of coma for one zone of the lens. These are the three transverse aberrations, the contributions of the front component being equal and opposite to the contributions of the rear. The two half-systems also contribute identical amounts to each of the longitudinal aberrations, but now the contributions have the same sign and add up instead of cancelling out.

The reason for this cancellation of the transverse aberrations can be seen by consideration of Fig. 108a. Any principal ray in any wavelength starting

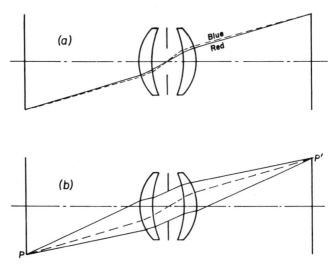

FIG. 108. Transverse aberrations of a holosymmetrical system. (a) Distortion and lateral color. (b) Coma.

out from the center of the stop and traveling both ways to the object and image planes will intersect those planes at the same height above and below the axis, giving a magnification of exactly -1.0 over the entire field. Thus distortion and lateral color are automatically absent. To demonstrate the absence of coma, we must trace a pair of upper and lower oblique rays in the stop both ways until they intersect each other at P and P' (Fig. 108b). We then add a principal ray through the center of the stop at such a slope that it passes through P. Symmetry will then dictate that it will also pass exactly through P'. Thus this one zone of the lens will be coma-free, although one cannot draw any similar conclusion for other zones of the lens. It should be noted that if there is any coma in each half of the lens, the principal ray in the stop will not be parallel to the parallel upper and lower oblique rays initially placed there.

If the lens is symmetrical but the conjugates are not equal, then the distortion will be corrected only if the entrance and exit pupils, where the entering and emerging portions of the principal ray cross the axis, are fixed points for all possible obliquity angles. Similarly, lateral color will be absent if the entrance and exit pupils are fixed points for all wavelengths of light. These two conditions are often referred to as the Bow–Sutton condition. No corresponding conclusions can be drawn for coma, but it is generally found that coma is greatly reduced by symmetry, even though the conjugate distances are not equal. The point to notice is that if distortion and lateral color must be well corrected over a wide range of magnifications, as in a process lens used to copy maps, then the designer must concentrate on correcting the spherical and chromatic aberrations of the principal rays rather than on the primary image, stopping the lens down if necessary to maintain the image quality. Stopping the lens down, of course, has no effect on the aberrations of the principal ray.

VI. COMPUTATION OF THE SEIDEL ABERRATIONS

In some designs it is advantageous to determine the contributions of the various surfaces, or thin lens elements, to the seven primary or Seidel aberrations. This procedure has the advantage of indicating where each aberration arises in the system, and the computation is rapid enough to permit an approximate design to be reached in a short time before any real ray tracing is attempted.

To calculate the surface contributions, we first trace a regular paraxial ray from object to image, and also a paraxial principal ray through the center of the stop. The entering values of the (y, u) of the paraxial ray and the (y_{pr}, u_{pr}) of the paraxial principal ray must correspond to the desired values for the real lens, so that the y is equal to the true Y at the first surface, and the u_{pr} is equal to the tan U_{pr} of the angular field for which the primary aberrations are desired. In this notation, the Lagrange invariant can be written

$$hnu = n(y_{pr}u - u_{pr}y)$$

A. SURFACE CONTRIBUTIONS

Both Conrady[6] and Feder[7] have given simple formulas by which the surface contributions to the Seidel aberrations can be rapidly computed. We calculate the following equations in order, noting that subscript 0 in u'_0 and h'_0 refers to the final image, while other symbols refer to the surface in question. Having traced the paraxial ray and the paraxial principal ray, we calculate their angles of incidence by the usual relation $(i = yc - u)$, where c is the surface curvature. Then

[6] A. E. Conrady, pp. 314, 751.

[7] D. P. Feder, Optical calculations with automatic computing machinery, J. Opt. Soc. Am. **41**, 633 (1951).

$$K = yn\left(\frac{n}{n'} - 1\right)(i - u')/2u_0'^2$$

$$SC = Ki^2, \qquad CC = Kii_{pr}u_0', \qquad AC = Ki_{pr}^2$$

$$PC = -\frac{1}{2}h_0'^2 c\left(\frac{n' - n}{nn'}\right)$$

$$DC = (PC + AC)(u_0'i_{pr}/i) = CC_{pr} + \tfrac{1}{2}h_0'(u_{pr}'^2 - u_{pr}^2) \qquad (88)$$

$$L = yn\left(\frac{\Delta n}{n} - \frac{\Delta n'}{n'}\right)/u_0'^2$$

$$L_{ch}C = Li, \qquad T_{ch}C = Li_{pr}u_0'$$

Here c is the surface curvature, as usual; for the aberrations, SC is the contribution to longitudinal spherical aberration, CC to the sagittal coma, AC to sagittal astigmatism (i.e., the longitudinal distance from the Petzval surface to the sagittal focal line), PC to the sag of the Petzval surface, and DC to the distortion; $L_{ch}C$ and $T_{ch}C$ are the surface contributions to the longitudinal chromatic aberration and lateral color respectively. Of the two expressions for the distortion contribution, the first is easier to compute by hand, but it fails if the object is at the center of curvature of the surface, for then $AC = -PC$ and DC becomes indeterminate. The second alternative expression requires the calculation of CC for the principal ray, and it is therefore recommended that a subroutine be written for CC that can be applied successively to the paraxial ray and the principal ray data.

B. THIN-LENS CONTRIBUTIONS

In some thin-lens predesigns it is convenient to reduce the system to a succession of thin elements separated by finite air spaces. We trace the same two paraxial rays through the system, using Eqs. (26) for this purpose. We then list the values of $Q = (y_{pr}/y)$ at each thin lens. The computation now falls into two parts: first, the calculation of each contribution as if the stop were at the thin lens, and then the modification of each contribution to place the stop in its true position. The second stage makes use of the Q at each lens.

The equations for the first stage are

$$SC = -\frac{y^4}{u_0'^2}\sum (G_1 c^3 - G_2 c^2 c_1 + G_3 c^2 v_1$$

$$+ G_4 cc_1^2 - G_5 cc_1 v_1 + G_6 cv_1^2)$$

$$CC = -y^2 h_0' \sum \left(\tfrac{1}{4}G_5 cc_1 - G_7 cv_1 - G_8 c^2\right)$$

$$AC = -\tfrac{1}{2}h_0'^2 \sum (1/f) \qquad (89)$$

$$PC = -\tfrac{1}{2}h_0'^2 \sum (1/nf)$$

$$DC = 0, \qquad T_{ch}C = 0$$

$$L_{ch}C = -\frac{y^2}{u_0'^2} \sum \left(\frac{1}{Vf}\right)$$

The summations in these expressions are used only if the thin component is compound, such as a thin doublet or a thin triplet; they are not required for a single thin element.

We next apply the calculated Q factors ($Q = y_{pr}/y$) to place the stop into its correct position. The true contributions, marked with asterisks, are found in the following way:

$$SC^* = SC, \qquad PC^* = PC, \qquad L_{ch}C^* = L_{ch}C$$

$$CC^* = CC + SC(Qu_0')$$

$$AC^* = AC + CC(2Q/u_0') + SCQ^2 \qquad (90)$$

$$DC^* = (PC + 3AC)Qu_0' + 3CCQ^2 + SC(Q^3 u_0')$$

$$T_{ch}C^* = L_{ch}C(Qu_0')$$

These expressions are generally known as the stop–shift formulas.

C. ASPHERIC SURFACE CORRECTIONS

If we are computing the Seidel surface contributions and encounter an aspheric surface, we first calculate the contributions assuming that the surface is a sphere with the vertex curvature c, and then add a set of correcting terms depending on the asphericity.

The aspheric surface is assumed to be of the form

$$X = \tfrac{1}{2}cS^2 + j_4 S^4 + j_6 S^6 + \cdots$$

where $S^2 = x^2 + y^2$ and the j values are the aspheric coefficients. Then

$$\text{addition to } SC = 4j_4\left(\frac{n-n'}{u_0'^2}\right)y^4$$

$$\text{addition to } CC = 4j_4\left(\frac{n-n'}{u_0'^2}\right)y^3 y_{pr} \qquad (91)$$

$$\text{addition to } AC = 4j_4 \left(\frac{n - n'}{u_0'^2} \right) y^2 y_{\text{pr}}^2$$

$$\text{addition to } DC = 4j_4 \left(\frac{n - n'}{u_0'^2} \right) y y_{\text{pr}}^3$$

It should be noted that only the j_4 coefficient appears in the primary aberrations, since the higher aspheric terms affect only the higher-order aberrations. Also, if the stop is located at an aspheric surface, the y_{pr} there will be zero, and the only aberration to be affected by the asphericity is the spherical aberration.

D. A Thin Lens in the Plane of an Image

This case is exemplified by a field lens or a field flattener. We cannot now use the thin-lens contribution formulas already given because both the stop and the image cannot lie in the same plane. Consequently, we have to return to the surface contribution formulas and add them up for the case in which $y = 0$. When this is done, we find that for a thin lens situated in an image plane

$$SC = CC = AC = 0, \qquad L_{\text{ch}} C = T_{\text{ch}} C = 0$$

The Petzval sag PC has its usual value of $-\frac{1}{2} h_0'^2 / f' N$, where N is the index of the glass. The distortion must be carefully evaluated. It turns out to be

$$DC = \tfrac{1}{2} h_0'^2 u_0' \left(\frac{y_{\text{pr}}}{N f' u_1} \right) \left[\frac{1}{r_1} + \frac{N}{r_2} - \frac{1}{l_{\text{pr}1}} \right]$$

where f' is the focal length of the thin lens. The distortion contribution depends on the shape of the thin lens in addition to its focal length and refractive index.

CHAPTER 11

Lenses in Which Stop Position Is a Degree of Freedom

It is obvious that, depending on its position in a lens system, a stop selects some rays from an oblique pencil and rejects others. Thus, if the stop is moved along the axis (or for that matter, if it is displaced sideways, but that case will not be considered here), some of the former useful rays will be excluded while other previously rejected rays are now included in the image-forming beam. Consequently, unless the lens happens to be perfect, a longitudinal stop shift changes all the oblique aberrations in a lens. It will not affect the axial aberrations provided the aperture diameter is changed as necessary to maintain a constant F number.

I. THE $H' - L$ PLOT

The results of a stop shift can be readily studied by tracing a number of meridional rays at some given obliquity through the lens, and plotting a graph connecting the intersection length L of each ray from the front lens vertex as abscissa, with the intersection height H' of that ray at the paraxial image plane as ordinate. This graph (Fig. 109) is similar to the meridional

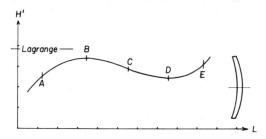

FIG. 109. A typical $H' - L$ graph for a meniscus lens.

ray plot discussed on p. 144, except that the abscissas have reversed sign, so that the upper rim ray of the beam now appears at the left-hand end of the graph while the lower rim ray falls at the right. Locating a stop in any position selects a portion of the graph and rejects the rest of it. The ray passing though the center of the stop is, of course, the principal ray of the useful beam.

This graph tells us a great deal about the aberrations in the image and how they will be changed when the stop is shifted along the axis:

1. Distortion

The height of the graph at the principal-ray point above or below the Lagrangian image height is a direct measure of the distortion.

2. Tangential Field Curvature

The slope of the graph at the principal-ray point is a measure of the sag of the tangential field X'_t, a quantity that is ordinarily determined by the Coddington equations. If the slope is upward from left to right it indicates an inward-curving field because the upper rim ray strikes the image plane lower than the lower rim-ray. A graph that is horizontal at the principal-ray point indicates a flat tangential field.

3. Coma

The curvature of the graph at the principal-ray point is a measure of the tangential coma present in the lens. If the ends of the portion of the graph used are above the principal ray this indicates positive coma. The coma is clearly zero at a point of inflection where the graph is momentarily a straight line.

4. Spherical Aberration

The presence of spherical aberration is indicated by a cubic or S-shaped curve, undercorrection giving a graph in which the line joining the ends of the curve is more uphill than the tangent line at the principal-ray point.

All of these phenomena are illustrated in the typical $H' - L$ graph shown in Fig. 109. If the principal ray falls at A, the field will be drastically inward-curving. At B the field is flat but there is strong negative coma. At C the coma is zero but the field is now backward-curving. At D the field is once more flat but now the coma is positive, while at E the field is once more drastically inward-curving. The overall S shape of the curve indicates the presence of considerable undercorrected spherical aberration.

Thus we reach the important conclusion that we can eliminate coma by a suitable choice of stop position if there is spherical aberration present; indeed, this result is implicit in the OSC formulas on p. 161. Furthermore, we can flatten the tangential field by a suitable choice of stop position if there is a sufficiently large amount of coma or spherical aberration or both. In terms of the primary or Seidel aberrations, these conclusions are in agreement with the stop–shift equations given on p. 207.

II. SIMPLE LANDSCAPE LENSES

It is instructive to plot the $20°$ $H' - L$ curves for a single lens bent into a variety of shapes, as in Fig. 110. In curve (a) the lens is shown bent into a

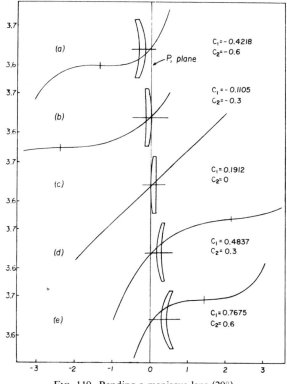

FIG. 110. Bending a meniscus lens ($20°$).

strongly meniscus shape, concave to the front where parallel light enters. There is a large amount of spherical undercorrection, leading to an S-shaped cubic curve, and the interesting region containing the maximum, inflection, and minimum points lies close to the lens. In these graphs the abscissa values are measured from the anterior principal point in each case. The focal length is everywhere 10.0, the thickness 0.15, and the refractive index 1.523.

In curve (b) the lens is bent into such a weak meniscus shape that there is very little spherical aberration, with no maxima or minima. In curve (c) is shown a plano-convex lens with curved face to the front. There is now no coma and very little spherical aberration so that the curve is practically a straight line. In the remaining graphs the lens is a meniscus with convex side to the front, and now the interesting region has moved behind the lens, still on the concave side of it. It will be noticed that all the graphs have about the same slope at $L = 0$. This bears out the well-known fact that any reasonably thin lens with stop in contact has a fixed amount of inward-curving field independent of the structure of the lens.

As a simple meniscus lens has only two degrees of freedom, namely, the lens bending and the stop position, it is clear that only two aberrations can be corrected. Invariably the two aberrations chosen are coma and tangential field curvature. The axial aberrations, spherical and chromatic, can be reduced as far as necessary by stopping the lens down to a small aperture, and $f/15$ is common although some cameras with short focal lengths have been opened up as far as $f/11$. The remaining aberrations, lateral color, distortion, and Petzval sum, must be tolerated since there is no way to correct them in such a simple lens. Changes in thickness and refractive index have very little effect on the aberrations.

In designing a landscape lens, therefore, we must choose a bending such that the $H' - L$ curve is a horizontal line at the inflection point. This will ensure that the coma is corrected and the tangential field will be flat at whatever field angle was chosen for plotting the $H' - L$ curve. The field may turn in or out at other obliquities, of course.

A. A SIMPLE REAR LANDSCAPE LENS

To meet the specified conditions, it is found by interpolating between the examples shown in Fig. 110 that for a simple rear landscape lens a front surface curvature of about -0.28 is required. With the thickness and refractive index used here there is very little latitude. Solving the rear curvature to give a focal length of 10.0, we arrive at the 25° $H' - L$ curve shown in

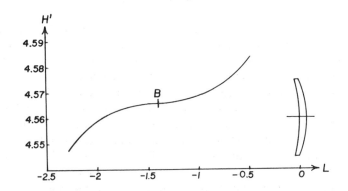

FIG. 111. The $H' - L$ graph of a rear meniscus lens having a flat coma-free field (25°).

Fig. 111. This curve indicates that the stop must be at B, a distance of 1.40 in front of the lens. At $f/15$ the stop diameter will be 0.667, and to cover a field of up to 30° the lens diameter must be about 1.80. Actually, because of excessive astigmatism, it is unlikely that this lens would be usable beyond about 25° from the axis. The lens system is

c	d	n
-0.28		
	0.15	1.523
-0.4645		

with $f' = 10.0003$, $l' = 10.1445$, LA' $(f/15) = -0.2725$, Petzval sum = 0.0634. The astigmatism is found to be (Fig. 112)

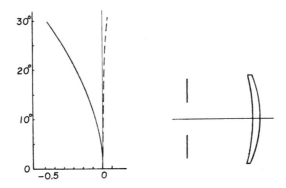

FIG. 112. Astigmatism of a simple rear-meniscus lens.

Field (deg)	X'_s	X'_t	Distortion (%)
30	-0.584	0.044	-3.34
20	-0.260	0.003	-1.41

If a flatter form were used, the spherical aberration would be slightly reduced and the tangential field would be inward-curving. This would reduce the astigmatism, but the sagittal field is already seriously inward-curving and flattening the lens would make it even worse. It therefore appears that the present design is about as good as could be expected with such a simple lens.

B. A SIMPLE FRONT LANDSCAPE LENS

Quite by chance, lens (e) in Fig. 110 has a horizontal inflection, and it therefore meets the requirements for a landscape lens. Its structure is

c	d	n
0.7675		
	0.15	1.523
0.60		

with $f' = 9.99918$, $l' = 9.60387$, $LA' (f/15) = -0.4729$, stop distance $= 0.8641$, stop diameter $= 0.5830$, Petzval sum $= 0.0575$. Then

Field (deg)	X'_s	X'_t	Distortion (%)
30	-0.603	$+0.074$	$+3.50$
20	-0.246	$+0.005$	$+1.41$

The lens diameter should be about 1.6 (Fig. 113). Note that the surfaces on

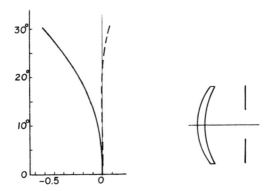

FIG. 113. Astigmatism of a front-meniscus lens.

this lens are much stronger than those of the rear meniscus and the aberration residuals are generally worse. The spherical aberration particularly is much greater than in the previous system. Nevertheless, front landscape lenses are usually preferred because the camera can then be made shorter for the same focal length, and the large front lens acts as an effective shield to prevent the entry of dirt into the shutter mechanism. In an effort to reduce the spherical aberration a flatter lens is often employed, but the resulting inward-curving field must then be offset in part by the use of a cylindrically curved film gate. The compensation is not very good, however, because a cylinder does not fit very well on a spherically curved image.

It is worth noting that the remaining spherical and chromatic aberration residuals in a simple landscape lens have the effect of greatly increasing the depth of field, so that if the film is correctly located relative to the lens, then any object from, say, 6 ft to infinity will be sharply imaged on the film in some particular wavelength and some particular lens zone, the other wavelengths and other zones being more or less out of focus. Thus we have a sharp image superposed on a slightly blurred image of objects at all distances (within limits), and if the exposure is kept on the short side, very acceptable photographs can be obtained without the necessity for any focusing mechanism on the camera.

III. A PERISCOPIC LENS

It was found empirically very early in the development of photography that placing two identical landscape lenses symmetrically about a central stop removed the distortion and lateral color, giving a better image than could be obtained by the use of a simple meniscus lens alone. Such a lens was called " periscopic."

To design the rear half of a symmetrical lens, we assume that there will be parallel light in the stop space, and now we can evidently ignore coma since it will be corrected automatically by the symmetry. Therefore we have to consider only the tangential field curvature, and by the $H' - L$ curve we can select a stop position to flatten the field provided the lens bending is equal to or stronger than that used for a landscape lens, but we cannot use stop position to flatten the field if the bending is weaker than that of a landscape lens. Also, the steeper the bending the closer will the stop be to the lens, resulting in a more compact system.

Using the thickness and refractive index employed in our previous designs, we will try a rear-meniscus lens with $c_1 = -0.8$. The structure is

c	d	n
-0.8		
	0.15	1.523
-0.95198		

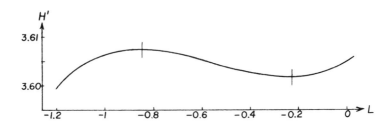

FIG. 114. The $H' - L$ curve of the rear component of a periscopic lens (20).

with $f' = 9.99975$, $l' = 10.41182$. The $H' - L$ curve for a $20°$ obliquity is shown in Fig. 114. This graph tells us that our lens will have a flat tangential field if the stop is placed at a distance of -0.85 or -0.23 from the front lens surface. Naturally, we choose the nearer position, and we mount two similar lenses about a central stop. The focal length now drops to 5.3874, and so we scale up the combined system to a focal length of 10. The system now is as shown in Fig. 115a:

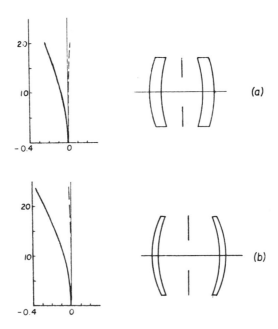

FIG. 115. Two periscopic designs.

c	d	n
0.51287		
	0.278	1.523
0.431		
	0.427	
	0.427	
−0.431		
	0.278	1.523
−0.51287		

with $f' = 10.00414$, $l' = 9.32841$, stop diameter $(f/15) = 0.620$, LA' $(f/15) =$ -0.2959, Petzval sum $= 0.0562$.

Field (deg)	X'_s	X'_t	Distortion (%)
19.8	−0.231	+0.019	+0.04

The spherical aberration and field curvature calculated here are for parallel light entering the left-hand end of the system, but it was designed on the assumption that there would be parallel light in the stop. It is actually rather surprising that the aberrations for a distant object resemble so closely the aberrations of the rear half alone. It is clear that the tangential field is slightly too far backward, and it is therefore desirable to reduce the central air space slightly to flatten the field. Also, the scaling-up process has made the lens elements unnecessarily thick, and it would be worth going back to the beginning and redesigning the system with much thinner lenses.

It is of interest to compare this design with the original Steinheil "Periskop" lens, which was of this type. According to von Rohr,[1] the specification was

c	d	n
0.5645		
	0.1316	1.5233
0.4749		
	0.6484	
	0.6484	
−0.4749		
	0.1316	1.5233
−0.5645		

[1] M. von Rohr, "Theorie und Geschichte des photographischen Objektivs," p. 288. Springer, Berlin, 1899.

with $f' = 10$, $l' = 9.2035$, stop diameter $(f/15) = 0.627$, LA' $(f/15) = -0.355$, Petzval sum $= 0.0615$.

Field (deg)	X'_s	X'_t	Distortion (%)
23.4	−0.364	−0.010	+0.07

The lens is shown in Fig. 115b.

IV. ACHROMATIC LANDSCAPE LENSES

A. THE CHEVALIER TYPE

In this type a flint-in-front lens of slightly meniscus shape is used, with stop in front and the concave side facing the distant object.

As an example we will use the following glasses:

(a) Schott: F-10, $n_d = 1.62360$, $V = 36.75$, $\Delta n = 0.01697$
(b) Schott: BK-13, $n_d = 1.52122$, $V = 62.72$, $\Delta n = 0.00831$

For a focal length of 10, we find

$$c_a = -0.2269, \qquad c_b = +0.4634$$

Assuming an equiconcave flint as a starter and establishing suitable thicknesses (actually those used here were too thick), we solve the last radius by the $D - d$ method and find the focal length to be 10.515. After scaling down to a focal length of 10 we have

c	d	n
−0.1189		
	0.28	1.62360
0.1189		
	0.56	1.52122
−0.3424		

with $f' = 10.00$, $l' = 10.4510$, LA' $(f/15) = -0.162$, Petzval sum $= 0.0667$.

We next trace a set of oblique rays through the upper half of the lens at $-20°$ to locate the stop position for zero coma (Fig. 116). This gives

L	H'
0	3.579278
−1.0	3.566382
−2.0	3.577830
−3.0	3.576493

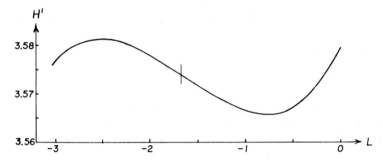

FIG. 116. The $H' - L$ curve of a Chevalier achromat (20°).

The inflection point of this graph is at $L = -1.67$, and because the graph is S-shaped, the tangential field will obviously be backward-curving. Coddington traces at several obliquities give (Fig. 117)

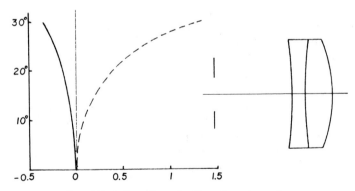

FIG. 117. Astigmatism of a Chevalier achromat.

Field (deg)	X'_s	X'_t	Distortion (%)
30	−0.341	1.325	−4.18
20	−0.134	0.394	−1.84
10	−0.043	0.079	−0.46

Unfortunately, it is not possible to flatten the tangential field in a lens of this type at the same time as eliminating the coma. The concave front face and the dispersive interface both contribute overcorrected astigmatism of about the same amount, and bending the lens merely increases one contribution while reducing the other. Using modern barium crown glass with a flint of the same index, one could make an achromatic lens that would behave like a simple landscape lens so far as the monochromatic aberrations are concerned, the interface then being merely a buried surface. Another possibility is to depart from strict achromatism by weakening the cemented interface, but the high cost of such a lens over that of a single element would be scarcely justified.

B. The Grubb Type

In 1857 Thomas Grubb[2] made a lens that he called the Aplanat, consisting of a meniscus-shaped crown-in-front achromat. The spherical aberration was virtually corrected by the strong cemented interface, and as a result the user had to accept either the coma or the field curvature since both could not be corrected together. The Grubb lens eventually led to the "Rapid Rectilinear" design discussed in Section V,A.

C. A "New Achromat" Landscape Lens

Since the cemented interface in the "old" Chevalier achromat has the effect of overcorrecting the astigmatism at high obliquities, it is evident that we could reverse the effect if we were to use a crown glass of higher refractive index than the flint glass (a "new achromat"). Furthermore, this combination of refractive indices has the effect of reducing the Petzval sum, but it will be accompanied by a large increase in the spherical aberration.

The design procedure for a new achromat is entirely different from that for an old achromat, because now we leave the achromatizing to the end and solve the outside radii of curvature for Petzval sum and focal length. We select refractive indices such that there is a variety of dispersive powers available for achromatizing after the design is completed. Two typical refractive indices meeting this requirement are

(a) Flint: 1.5348 (available V numbers from 45.7 to 48.7)
(b) Crown: 1.6156 (available V numbers from 54.9 to 58.8)

As a first guess we will aim at a Petzval sum of 0.03 on a focal length of 10. We must also guess at a likely interface radius and lens thicknesses. This gives as a starting system

---

[2] T. Grubb, Brit. Patent 2574/1857.

c	d	n
-0.551		
	0.1	1.5348
0.164		
	0.4	1.6156
-0.5687		

with $f' = 9.9998$, $l' = 10.8865$, Petzval sum $= 0.030$. The large thickness helps to reduce the Petzval sum without using very strong elements.

In plotting the $H' - L$ graph, we use a larger obliquity angle than before because new achromats tend to cover an exceptionally wide field. The graph shown in Fig. 118 represents the curve for $-25°$, and we see that the

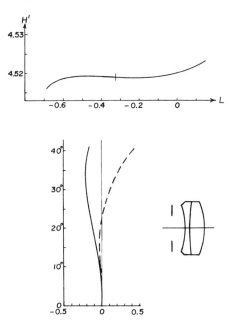

FIG. 118. Tentative design of a new-achromat lens.

inflection falls at $L = -0.326$. The astigmatic field curves are also shown.

It is evident that the collective interface should be made stronger to move the field inward, and a smaller Petzval sum would also be desirable. Hence for our next attempt we try $c_2 = 0.25$ and Petzval sum $= 0.027$. These changes give (Fig. 119)

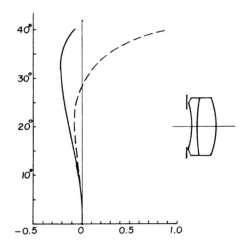

FIG. 119. Astigmatism of a new achromat, later form.

c	d	n
−0.5777		
	0.1	1.5348
0.25		
	0.4	1.6156
−0.57795		

with $f' = 9.99996$, $l' = 10.91516$, Petzval sum $= 0.027$, LA' $(f/15) = -0.50$, stop position $= -0.102$. The fields are

Field (deg)	X'_s	X'_t	Distortion (%)
40	−0.075	0.869	−9.23
30	−0.207	0.027	−4.68
20	−0.130	−0.083	−2.04
10	−0.041	−0.041	−0.46

Assuming that this is acceptable, the last step is the selection of real glasses for achromatism. A few trials, using the $D - d$ method, indicate that the following glasses would be excellent:

(a) LLF-8: $n_e = 1.53530$, $\Delta n = 0.01158$, $V = 46.23$
(b) SK-4: $n_e = 1.61521$, $\Delta n = 0.01046$, $V = 58.82$

It is perhaps not obvious which of these two designs would be the better. For a narrow field such as $\pm 22°$, the lens in Fig. 118 is to be preferred, while for a wider field such as $\pm 33°$ the lens in Fig. 119 would obviously be better. It is interesting to see how the small changes in the design have made such a large difference to the tangential field at the wider field angles.

The large spherical aberration is a definite disadvantage of the new achromat form. This was corrected by Paul Rudolph in his Protar design, to be discussed in Chapter 12.

V. ACHROMATIC DOUBLE LENSES

A. The Rapid Rectilinear

The Rapid Rectilinear, or Aplanat, lens is one of the most popular photographic lenses ever made. The lens is symmetrical, and the rear half is spherically corrected and has a flat field. In order to keep the lens compact, a large amount of positive coma is required in the rear component. This implies that a graph of spherical aberration against bending should rise high above the zero line, much higher than is usual for telescope objectives. To achieve this, the V difference between the old-type crown and flint glasses should be small, but a large index difference is helpful. The exact V difference depends on the aperture and field required. For a normal lens of $f/6$ or $f/8$ aperture, a V difference of about 7.0 is satisfactory. A smaller V difference can be used for a wide-angle lens of $f/16$ aperture, while a larger V difference leads to a longer lens of higher aperture, suitable for portraiture. All three of these variations have been used by different manufacturers. At first, two flint glasses were utilized, but after about 1890 it was common to find an ordinary crown in combination with a light barium flint.

To indicate the design procedure, we will select the following glasses:

(a) LF-1: $n_e = 1.57628$, $\Delta n = n_F - n_C = 0.01343$, $V = 42.91$
(b) F-1: $n_e = 1.63003$, $\Delta n = n_F - n_C = 0.01756$, $V = 35.87$

$V_a - V_b = 7.04$. In designing the rear component, the procedure already described for telescope doublets is followed, except that because of the strongly meniscus shape of the lenses, the preliminary G sum analysis is not very helpful and will be omitted.

Using these glasses, the (c_a, c_b) formulas give, for a focal length of 10,

$$c_a = 1.0577, \qquad c_b = -0.8089$$

Assuming that c_1 will be about one-half c_a with negative sign, we make a drawing of the lens at a diameter of about one-tenth the focal length, enabling us to set the thicknesses at 0.3 for the crown and 0.1 for the flint.

Taking a few bendings and solving each for perfect achromatism by the $D - d$ method on a traced $f/16$ ray, we can plot the graph in Fig. 120,

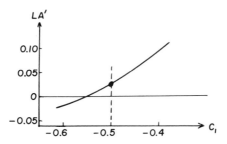

FIG. 120. Bending curve for the rear component of a Rapid Rectilinear.

connecting the spherical aberration with c_1 in the neighborhood of the left-hand solution having positive coma. The right-hand solution with negative coma is useless since it would require the stop to be behind the lens to flatten the field. Since this is a photographic lens, we desire a small amount of spherical overcorrection to offset the zonal undercorrection, which suggests that we try $c_1 = -0.5$ for further study. This lens has a focal length of 10.806, $LA'_m = +0.026$, and $LZA = -0.0178$. To find the stop position for a flat tangential field, we plot the $H' - L$ graph at 20° for a succession of L values as in Fig. 121b. The minimum point falls at $L = -0.2$, which is the

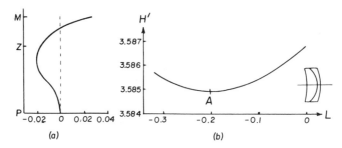

FIG. 121. Aberrations of the rear component of a Rapid Rectilinear. (a) Spherical aberration, (b) the $H' - L$ curve at 20° obliquity.

distance from the stop to the front (concave) surface.

We now assemble two of these lenses together about a central stop, and find that the focal length is 5.6676. It is best to scale this immediately to a focal length of 10.0, giving the following specification (Fig. 122a):

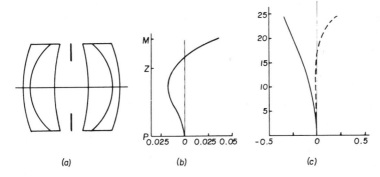

FIG. 122. The final Rapid Rectilinear design.

c	d	n
0.3974		
	0.1764	1.63003
0.8828		
	0.5293	1.57628
0.2834		
	0.3529	
	0.3529	
−0.2834		
	0.5293	1.57628
−0.8828		
	0.1764	1.63003
−0.3974		

with $f' = 10.00$, $l' = 9.0658$, lens diameter $= 1.8$, Petzval sum $= 0.0630$. The $f/8$ axial ray from infinity gives $LA' = 0.0350$, and it also tells us that the $f/8$ stop diameter must be 1.110. An $f/11.3$ zonal ray gives $LZA = -0.0108$, enabling us to plot the spherical aberration graph in Fig. 122b.

To plot the fields, we now add two other principal rays having slope angles in the stop space of 28 and 12°, respectively. The sagittal and tangential fields traced along these principal rays are as follows:

Outside angle (deg)	Angle at stop (deg)	X'_s	X'_t	Distortion (%)
24.4	28	−0.3411	0.2050	0.09
17.5	20	−0.2013	0.0044	0.04
10.6	12	−0.0789	−0.0196	0.01

The fields are plotted in Fig. 122c and closely resemble those of the rear half system. Both the spherical aberration and the astigmatism are thus very stable in this type of lens for changes in the object distance, which was one of the reasons for its great popularity.

B. A FLINT-IN-FRONT SYMMETRICAL ACHROMATIC DOUBLET

There is, of course, a companion system to the Rapid Rectilinear in which the rear component is a flint-in-front spherically corrected achromat. To design such a lens we may use the same glasses as for the Rapid Rectilinear, and we plot a graph of spherical aberration at $f/16$ against bending, of course in the region of the left-hand solution where the coma is positive (Fig. 123).

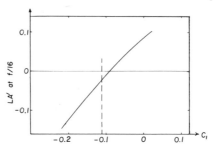

FIG. 123. Spherical aberration vs. bending for a flint-in-front doublet.

For each plotted point the last radius is solved for strict achromatism by the $D - d$ method, and the curvatures are scaled to a focal length of 10, keeping the thicknesses at 0.1 and 0.3 as before.

We recall that when we were designing a telescope objective, we found that the left-hand solution for a flint-in-front doublet has a much smaller zonal residual than the left-hand crown-in-front doublet. Consequently we shall plan the present design to be a "portrait" lens with an aperture of $f/4.5$ and covering a somewhat narrower field than the rapid rectilinear. The rear half of the new lens will therefore have to work at $f/9$, and since the graph in Fig. 123 represents the $f/16$ aberration, we must select a bending having a small residual of undercorrected aberration, at say $c_1 = -0.11$. This gives the following rear half system:

	c	d	n
	-0.11		
		0.1	1.63003
	0.69		
		0.3	1.57628
$(D - d)$	-0.3489		

with $f' = 10.0542$, $l' = 10.3008$, Petzval sum $= 0.0706$, LA' $(f/9) = -0.0336$, LA' $(f/11.4) = -0.0365$, LA' $(f/16) = -0.0254$. The residual aberration at $f/9$ was deliberately made negative since it was found that mounting two similar components about a central stop tended to overcorrect the aberration. The last radius was determined, of course, by the $D - d$ method as usual.

To locate the stop, we trace several rays at $-20°$, giving the $H' - L$ curve shown in Fig. 124. The minimum falls at $L = -0.50$ for a flat tangential

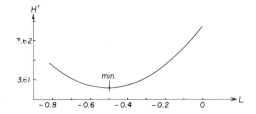

FIG. 124. The $H' - L$ graph of the rear component of a flint-in-front double lens (20°).

field. Mounting two of these lenses about a central stop and scaling to $f' = 10$ gives the following (Fig. 125):

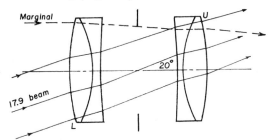

FIG. 125. Completed $f/4.5$ symmetrical portrait lens.

c	d	n
0.19450		
	0.5382	1.57628
−0.38462		
	0.1794	1.63003
0.06132		
	0.8970	
	0.8970	
−0.06132		
	0.1794	1.63003
0.38462		
	0.5382	1.57628
−0.19450		

with $f' = 10.0$, $l' = 8.4795$, Petzval sum $= 0.0787$, LA' $(f/4.5) = +0.0181$, LA' $(f/5.6) = -0.0069$. Then

Angle in object space	Angle in stop	X'_s	X'_t	Distortion (%)
24.956	28	−0.496	+0.543	+0.21
17.925	20	−0.294	−0.021	+0.10
10.798	12	−0.115	−0.055	+0.03

Plotting the fields and aberration of this lens makes an interesting comparison with the comparable data for the Rapid Rectilinear (Fig. 126). The

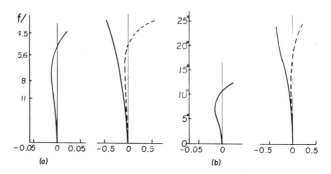

FIG. 126. Comparison of (a) flint-in-front and (b) crown-in-front (RR) forms. (Spherical aberration and astigmatism curves for $f' = 10$.)

reasons for regarding this as a portrait lens are evident here.

As a final check, we will trace a family of rays at $17.925°$ to complement the $20°$ principal ray already traced, and we plot the $(H - \tan U)$ curve shown in Fig. 127. The ends of this curve represent rays passing through the extreme top and bottom of the diaphragm, and as can be seen, the lower ray is very bad and should be vignetted off. It is customary in lenses of this kind to limit every surface to a clear aperture equal to the entering aperture of the marginal ray, which in this case is $Y = 1.1111$. This limitation cuts off the lower rays drastically, placing the true lower rim ray at the point marked L on the graph. It also somewhat reduces the upper part of the aperture to a limiting rim ray marked U.

It is clear that the remaining aberration of the lens is a small residual of negative coma of magnitude

$$\text{Coma}_t = \tfrac{1}{2}(H'_U + H'_L) - H'_{\text{pr}} = -0.0182$$

Assuming that the sagittal coma is one-third the tangential coma, the equi-

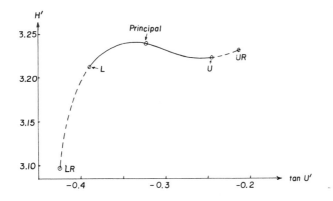

FIG. 127. The meridional ray plot for the final system at 17.9° field angle.

valent OSC becomes −0.00096, which is small enough to be neglected, especially since a lens of this type is unlikely to be used at a field as wide as 17.9°.

Although we have regarded this as a portrait lens, it has seldom been used in this way, but it could very well be used at or near unit magnification as a relay lens in a telescope, in which case the coma would automatically vanish.

C. LONG TELESCOPIC RELAY LENSES

In many types of telescope and periscope, an erector system working at or near unit magnification is inserted between the objective and eyepiece to give an erect image. This erector often consists of two identical spherically-corrected doublets mounted symmetrically about a central stop, the stop position being chosen to give a flat field exactly as in the Rapid Rectilinear lens, except that now we often need a long system rather than a short one. As was pointed out in connection with the design of the Rapid Rectilinear, the greater the amount of coma in the rear lens the smaller the stop shift required to give a flat tangential field, and the shorter the relay will be. For a long relay, therefore, we need a spherically corrected lens with a very small amount of coma, and hence we select a design in which the graph of spherical aberration against bending rises only a little way above the abscissa line. Furthermore, whatever lens we use for the rear component of our relay must have positive coma in order that the flat-field stop position will be in front of the lens.

Referring to the bending curve on p. 165 for a normal cemented doublet, we see that the left-hand solution has positive coma, and it is therefore suitable for the rear of a telescopic relay. We locate the stop position for a flat field as we did for the Rapid Rectilinear by tracing several oblique parallel rays through the upper half of the lens, a suitable obliquity being now about 4°. The left-hand flint-in-front solution is much preferable to the crown-in-front form since it has only about a third of the zonal aberration, and we will follow up that design here. The graph connecting H' with L for this lens is shown in Fig. 128, and since the minimum falls at $L = -3.2$, that will be the

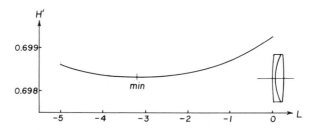

FIG. 128. The $H' - L$ curve of a flint-in-front telescope objective at 4° obliquity.

stop position in this case. The computed astigmatism at 4° when the stop is at that position is found to be $X'_s = -0.0117$ and $X'_t = +0.0006$, representing the desired flat tangential field. Two of these lenses mounted together about a central stop would therefore make an excellent relay (Fig. 129).

FIG. 129. A 1 : 1 telescopic relay consisting of two flint-in-front objectives with a stop at the flat-field position.

If a still longer relay is required, the spherical aberration graph must be lowered still further, and the left-hand solution for the near aplanat discussed on p. 172 can be used. In this case the stop position, calculated at a very small obliquity such as 2°, falls at a distance of 9.2 in front of the lens, close to the anterior focal point, making the system telecentric in the image space. The combination of two such systems forms a 1 : 1 afocal telecentric relay, which has been used in contour projectors to give a longer working distance, and in borescopes, where up to four relays can be assembled in sequence without any need for field lenses at the intermediate real images.

The main relay in a submarine periscope consists of a pair of highly corrected aplanatic objectives spaced apart by a distance equal to two or three times their focal length, the field angle being then less than 1°. In this case the astigmatism is negligible so long as the tangential field is flat. Coma is corrected by the symmetry, in the usual way.

D. The Ross "Concentric" Lens

This is the classic example of a symmetrical objective consisting of two deeply curved new achromats surrounding a central stop. It was patented by Schroeder in 1889, with the structure of the rear half, after scaling to a focal length of 10, according to von Rohr[3] as follows:

c	d	n	V
	0.194	(air)	
−1.94125			
	0.020	1.5366	48.69
0			
	0.071	1.6040	55.31
−1.78358			

with $f' = 10$, $l' = 10.5961$, $L_{pp} = +0.6166$, Petzval sum $= -0.00618$. The glasses are assumed to be light flint No. 26 and dense barium crown No. 20 in Schott's catalog of 1886. Tracing a fan of rays entering at $-20°$ gives the $H' - L$ curve shown in Fig. 130. Clearly the stop position for a flat tangen-

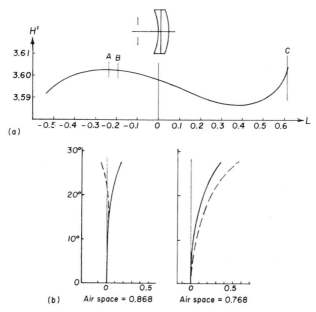

Fig. 130. Ross Concentric lens. (a) $H' - L$ curve of rear half (20°), (b) astigmatism of complete lens.

[3] M. von Rohr, "Theorie und Geschichte des photographischen Objektivs," p. 234. Springer, Berlin, 1899. Also U.S. Patent 404,506.

tial field should be at about $L = -0.237$ (point A), but von Rohr's specification places it at B, where $L = -0.194$, resulting in a slightly backward-curving field. (Incidentally, measurements made on an actual Concentric lens did not agree with this specification in any respect.) The front principal point is at C.

After combining two of these lenses together and scaling to a focal length of 10, the spherical aberration at $f/15$ was found to be -0.65, so that the lens should not be used at any aperture greater than about $f/20$. The fields with the preferred air space, and with that given by von Rohr, are also shown in Fig. 130. The unusual backward-curving sagittal field is, of course, due to the Petzval sum being negative. It is remarkable how great an effect a small change in the central air space has on the two fields.

CHAPTER 12

Symmetrical Double Anastigmats
with Fixed Stop

I. THE DESIGN OF A DAGOR LENS

For 25 years after the introduction of the Rapid Rectilinear lens, designers tried unsuccessfully to modify it in such a way as to reduce the Petzval sum, and so remove the astigmatism that limited the performance of the Rectilinear in the outer parts of the field. In 1892 the German designer von Höegh[1] made three useful suggestions to this end: (1) to insert a collective interface convex to the stop in the flint element of the Rapid Rectilinear, thus turning the half-system from a doublet into a triplet; (2) to use progressively increasing refractive indices outward from the stop; and (3) to use almost equal outside radii of curvature and to thicken the lens sufficiently to give the desired focal length and Petzval sum. In this way he created the famous "Double Anastigmat Goerz,"[2] later renamed the Dagor, which covered a fairly wide anastigmatic field at $f/6.3$. The symmetry, of course, automatically eliminated the three transverse aberrations, leaving the designer only spherical and chromatic aberration and astigmatism to be corrected in each half.

As an example in the design of such a lens, we first select three refractive indices for which there are many glasses available having different dispersive powers, so that we can achromatize the lens at the end by choosing suitable types of glass from available catalogues. These indices will be 1.517, 1.547, and 1.617, although of course other values could have been chosen that would be equally satisfactory. For a focal length of 10.0, we can start the design of the rear half-system with the radii -1, -0.5, $+2$, and -1, suitable thicknesses being determined from a scale drawing as 0.14, 0.06, and 0.19, respectively. These thicknesses are actually somewhat meager, but it is better to keep the lens as thin as possible to reduce vignetting at high field angles. Since the stop position will not be a degree of freedom, because we can correct all the aberrations without its help, we place the stop as close as possible to the lens, at a distance of 0.125.

---

[1] C. P. Goerz and E. von Höegh, U.S. Patent 528,155, filed Feb. 1893.
[2] A light-hearted account of the creation of the Dagor has been given by R. Schwalberg, *Pop. Phot.* **70**, 56 (1972).

We shall employ the four radii in the rear component in the following way: We vary r_1 to give the desired Petzval sum after solving r_4 for focal length. The two internal surfaces contribute very little to the sum. We find that a suitable value for the Petzval sum in a lens of this type is about 0.018 on a focal length of 10. We vary r_2 to make the marginal spherical aberration approximately equal and opposite to the 0.7 zonal aberration; and we select r_3 to give a flat tangential field on a principal ray passing through the center of the stop at an angle of 30°.

With the tentative initial data given above, we find that for the focal length and Petzval sum desired, the c_1 must be -0.78 and the $c_4 - 0.7748$. Our starting system is therefore as follows:

c	d	n
	0.125	(air)
-0.78		
	0.14	1.517
-2.00		
	0.06	1.547
0.50		
	0.19	1.617
-0.7748		

with $f' = 10.0$, $l' = 11.057$, Petzval sum $= 0.0182$, $f/12.5$ stop diameter $= 0.8$, LA' $(f/12.5) = 0.160$, LZA $(f/17.7) = -0.150$, X'_t at 30° $= -0.076$. A scale drawing of the lens is included in Fig. 131. By chance the spherical aberra-

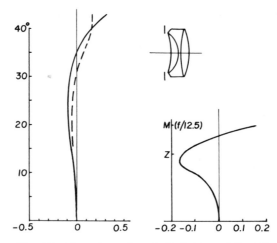

FIG. 131. Aberrations of rear half of a Dagor $(f' = 10)$.

tion is about right and will be accepted. However, the 30° tangential field is more inward-curving than we would like, so we proceed to reduce c_3 slightly, a suitable value being 0.486. This gives $X_t' = -0.0056$ and $X_s' = -0.0686$. Tracing a few more principal rays at other obliquities enables us to plot the field curves for the rear half of the system in Fig. 131.

Regarding this as a satisfactory rear half, we now assemble two of these lenses together and scale up to an overall focal length of 10.0. This gives the following for the front half:

c	d	n
0.4464		
	0.3304	1.617
−0.2795		
	0.1043	1.547
1.1502		
	0.2434	1.517
0.4486		
	0.2173	
	(symmetrical)	

with $f' = 10.0$, $l' = 9.2260$, Petzval sum $= 0.0211$, $f/6.8$ stop diameter $=$ 1.276, LA' $(f/6.8) = 0.0001$, LZA $(f/9.6) = -0.1130$. The spherical aberration is much less overcorrected than for the half-system, and the strong interfaces could be slightly deepened to rectify this. It will be noticed that the zonal aberration here is greater than in the corresponding Rapid Rectilinear lens; this is the major problem in lenses of the Dagor type. The fields are shown plotted in Fig. 132, and it will be seen that they are not greatly

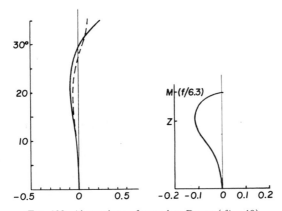

FIG. 132. Aberrations of complete Dagor ($f' = 10$).

different from those of the half-system. There is a minute amount of distortion, about 0.13% at 30°, which can be ignored.

The final step in the design is the selection of glasses for achromatism. We return to the marginal ray trace through the rear half-system, and compute the value of $D - d$ in each of the three lens elements. At an aperture of $f/12.5$ these are, respectively, -0.27299, $+0.50191$, and -0.25560. Our problem is therefore to find three glasses with the approximate indices used in this design, having Δn values such that $\sum (D - d) \Delta n = 0$. A brief search in the Schott catalog suggests that the following would make an achromatic combination:

(a) KF-3: $n_e = 1.51678$, $\Delta n = 0.00950$, $V_e = 54.40$
(b) KF-1: $n_e = 1.54294$, $\Delta n = 0.01079$, $V_e = 50.65$
(c) SK-6: $n_e = 1.61635$, $\Delta n = 0.01100$, $V_e = 56.08$

The refractive indices are close but not exactly equal to those assumed in the design. It would therefore be necessary to repeat the whole procedure using the exact index data for the real glasses, to obtain the final formula.

II. THE DESIGN OF AN AIR-SPACED DIALYTE LENS

This name is given to symmetrical systems containing four separated lens elements, as in Fig. 133. This type was originated by von Höegh,[3] who called

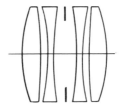

FIG. 133. The dialyte objective.

it the "Double Anastigmat Goerz type B," this name being later changed to Celor. The rear separated achromat contains five degrees of freedom, namely, two powers, two bendings, and an air space, with which it is possible to obtain the desired focal length and correct four aberrations: Petzval sum, spherical, chromatic, and astigmatism. If we then mount two of these components symmetrically about a central stop, we shall correct in addition the three transverse aberrations: coma, distortion, and lateral color. The stop

---

[3] C. P. Goerz and E. von Höegh, U.S. Patent 635,472, filed July 1898.

position is not a degree of freedom since we have sufficient variables without it. However, as the lens will generally be used with a distant object, we may have to depart slightly from perfect symmetry to remove any residuals of coma that may appear.

We can save a good deal of time by first determining the two powers and the separation of the rear component to yield the desired lens power, chromatic, and Petzval values, assuming thin lenses and using the Seidel contribution formulas given on p. 206. The thin-lens predesign requires the solution of the following three equations for ϕ_a, ϕ_b, and d:

$$\sum y\phi = y_a\phi_a + y_b\phi_b = \Phi y_a \qquad \text{(power)} \qquad (1)$$

$$\sum \frac{y^2\phi}{V} = \left(\frac{y_a^2}{V_a}\right)\phi_a + \left(\frac{y_b^2}{V_b}\right)\phi_b = -L'_{\text{ch}}u_0'^2 \qquad \text{(chromatic)} \qquad (2)$$

$$\sum \frac{\phi}{n} = \left(\frac{1}{n_a}\right)\phi_a + \left(\frac{1}{n_b}\right)\phi_b = \text{Ptz} \qquad \text{(Petzval)} \qquad (3)$$

From (1) we express y_b as a function of y_a, by $y_b = y_a(\Phi - \phi_a)/\phi_b$. Inserting this in (2) gives

$$\frac{y_a^2\phi_a}{V_a} + \frac{y_a^2(\Phi - \phi_a)^2}{\phi_b V_b} = -L'_{\text{ch}}u_0'^2 \qquad (4)$$

However, by (3),

$$\phi_b = n_b(\text{Ptz} - \phi_a/n_a) \qquad (5)$$

and putting this into (4) gives a quadratic for ϕ_a:

$$\phi_a^2[V_a - V_b n_b/n_a] + \phi_a[\text{Ptz } n_b V_b - 2\Phi V_a - L'_{\text{ch}}\Phi^2 V_a V_b n_b/n_a]$$
$$+ \Phi^2 V_a[1 + L'_{\text{ch}} \text{ Ptz } n_b V_b] = 0 \qquad (6)$$

Thus we obtain ϕ_a by (6), ϕ_b by (5), and finally the separation d by

$$d = (\phi_a + \phi_b - \Phi)/\phi_a\phi_b \qquad (7)$$

As an example in the design of such a lens, we will first solve the rear component for a focal length of 10, a Petzval sum of 0.030, and zero chromatic aberration. The selection of glasses must, however, be made with some care for if the V difference is too great the negative lens will be too weak to enable us to correct the other aberrations. A reasonable glass choice is

(a) Barium flint: $n_D = 1.6053$, $n_F = 1.61518$, $n_C = 1.60130$, $V = 43.61$
(b) Dense barium crown: $n_D = 1.6109$, $n_F = 1.61843$, $n_C = 1.60775$, $V = 57.20$

with V difference = 13.59. The above algebraic solution gives

$$\phi_a = -0.4958, \qquad \phi_b = 0.5458$$

and since $\phi = c/(n-1)$, we find that

$$c_a = -0.8191, \qquad c_b = 0.8934, \qquad d = 0.1848$$

This completes the predesign of the thin-lens powers and separation.

We could, of course, determine the bendings of the two thin elements in the rear half-system for spherical and astigmatic correction, but it is best to insert thicknesses first and assemble the two components before doing this. To assign thickness we must decide on the relative aperture of the finished lens, and $f/6$ is a good value to adopt. This makes the aperture of the rear component about $f/12$, and a diameter of 1.0 is suitable. The thicknesses of the two lenses will then be 0.06 for the flint and 0.20 for the crown.

For our starting bendings we may assign 40% of the total flint curvature to the front face of the rear negative element, and 25% of the crown curvature to the front face of the rear positive element. However, because of the finite thicknesses, we must scale each thick element to restore its ideal thin-lens power, and the air space must be adjusted to maintain the ideal separation between adjacent principal points. With the stop at a distance of 0.12 from the flint element, the whole lens becomes

c	d	r
$c_1 = -c_8 = 0.6788$		(1.473)
	0.2	
$c_2 = -c_7 = -0.2263$		(−4.418)
	0.0756	
$c_3 = -c_6 = -0.4893$		(−2.043)
	0.06	
$c_4 = -c_5 = 0.3262$		(3.065)
	0.12	
	(stop)	

The lens in its present state is drawn to scale in Fig. 133. The focal length is 5.6496 and the Petzval sum (for $f' = 10$) is 0.04039. Tracing $f/8.5$ rays in F and C light, with $Y_1 = 0.3323$, gives the zonal chromatic aberration as 0.03312. The increase in Petzval sum is due to the finite thicknesses of the lenses.

We must now restore the desired values of Petzval sum and chromatic aberration by changing the power of the two crown elements and the two outer air spaces, maintaining symmetry about the stop and letting the focal length go. Of course we could equally easily vary the power of the flint elements, but we must adopt a fixed procedure or we shall never reach a satisfactory solution. A double graph is a great convenience here, plotting the zonal chromatic aberration as ordinate and the Petzval sum for $f' = 10$ as abscissa. The starting point will be (0.0331,0.0404) and the aim point will be (0,0.034). A trial change of the outer air spaces by 0.05 gives chr. ab. $= -0.0016$ and Ptz $= 0.0368$, and then weakening both surfaces of the crown elements by 2% gives chr. ab. $= 0.0133$ and Ptz $= 0.0331$. The graph shown in Fig. 134 tells us that we should have increased the original air spaces by

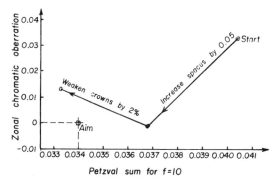

FIG. 134. Double graph for chromatic aberration and Petzval sum.

0.061 and weakened the crown lens surfaces by 1.24%. These changes give zero chromatic aberration and Petzval sum $= 0.0339$.

At this stage we find

$$LA' = \text{marginal spherical aberration} = 0.1335, \qquad X'_s \text{ at } 22° = 0.1088$$

$$LZA = \text{zonal spherical aberration} = 0.0429, \qquad X'_t \text{ at } 22° = 0.4216$$

We desire to have the marginal and zonal spherical aberrations equal and opposite, or $LA' + LZA = 0$, and we would also like to have $X'_t = 0$ for a flat tangential field. We proceed to accomplish this by bending both crowns and both flints in such a way as to maintain symmetry. In the double graph of Fig. 135, we see that at the start $X'_t = 0.4216$ and $LA' + LZA = 0.1764$. The aim point is (0, 0). Bending the crowns by $\Delta c_1 = -0.02$ toward a more nearly equiconvex form gives $X'_t = 0.1344$ and $LA' + LZA = 0.1742$. Then bending the flints by $\Delta c_3 = +0.02$ toward a more nearly equiconcave form gives $X'_t = 0.0254$ and $LA' + LZA = 0.0494$. The graph indicates that we should have used $\Delta c_1 = -0.0190$ and $\Delta c_3 = +0.0282$. These changes gave $X'_t = -0.0043$ and $LA' + LZA = 0.0022$, both of which are acceptable. The

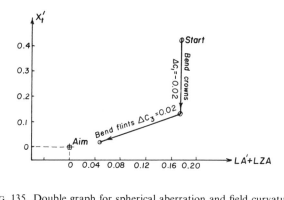

FIG. 135. Double graph for spherical aberration and field curvature.

Petzval sum for $f' = 10$ is now 0.0341 and the zonal chromatic aberration -0.0010; hence both are virtually unaffected by the small bendings that we have applied to the lenses.

At this point the symmetrical system has the construction

	c	d
$c_1 = -c_8 = 0.6514$		
		0.20
$c_2 = -c_7 = -0.2425$		
		0.1366
$c_3 = -c_6 = -0.4611$		
		0.06
$c_4 = -c_5 = 0.3544$		
		0.12
(stop)		

The four longitudinal aberrations are at the desired values, and we must now investigate the transverse aberrations to see how well they have been removed by the lens symmetry.

Tracing the $22°$ principal rays in C, D, and F light tells us that the distortion is 0.474% and the transverse chromatic aberration 0.000362. Since these are both positive, we can improve both at once by shifting a small amount of power from the front to the back. Weakening both surfaces of the front crown element by 2% and strengthening the rear crown surfaces by 2% lowers the distortion to 0.190% and reduces the transverse chromatic aberration to -0.00017, both of which are now acceptable. However, this change has slightly affected the other corrections, which are now

$$\text{focal length} = 5.4122$$

$$\text{zonal chromatic aberration} = -0.00022$$

$$\text{Petzval sum} = 0.0341, \quad \text{for} \quad f' = 10$$

$$LA' + LZA = -0.0158$$

$$X'_s = 0, \qquad X'_t = 0.0304$$

Since the last change has slightly altered the spherical aberration and field curvature, we return to the graph in Fig. 135 and apply $\Delta c_1 = -0.0034$ and $\Delta c_3 = -0.0027$ to restore these. The system is now as follows:

c	d	n_D
0.6350		
	0.2	1.6109
−0.2411		
	0.1366	
−0.4638		
	0.06	1.6053
0.3517		
	0.12	
	0.12	
−0.3517		
	0.06	1.6053
0.4638		
	0.1366	
0.2508		
	0.2	1.6109
−0.6610		

with focal length = 5.4212, zonal chromatic aberration = 0.00011, Petzval sum $(f' = 10) = 0.0342$, $LA' + LZA = -0.0022$; for 22°: $X'_s = -0.0088$, $X'_t = -0.0005$, distortion = 0.189%, lateral color = −0.00017; stop diameter for $f/6 = 0.7896$. Everything is thus known except coma, which we must now investigate.

The easiest way to evaluate the coma is to trace several oblique rays and draw the meridional ray plots at two or three obliquities and look for a parabolic trend, although in general this will be mixed with a cubic tendency due to oblique spherical aberration, and a general slope caused by inward or backward tangential field curvature. If the parabolic trend is not particularly noticeable, the amount of coma is probably negligible in view of the other aberration residuals that are unavoidably present. However, if coma is the dominant aberration it is necessary to reduce it by bending the two crown elements in the same direction, and not symmetrically about the stop as was done previously to correct the spherical aberration and field curvature.

In the present example, meridional ray fans were traced at the three obliquities 10, 16, and 22°, as shown in Fig. 136. The abscissas are the A

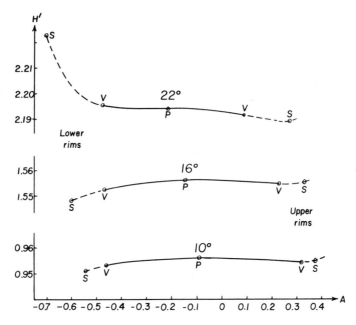

FIG. 136. Meridional ray plots for dialyte ($f' = 5.42$).

values of the rays, i.e., the height of incidence of each ray at the tangent plane to the front surface. To locate the endpoints of the curves we must decide what clear aperture we shall allow at the front and rear of the complete lens. It is customary in a short lens such as this to give all eight surfaces the initial aperture of the marginal beam, which in our case is $f/6$ or 0.904. The limiting rays at each obliquity are then found by trial such that the lower rim rays meet r_1 at a height of -0.452, and the upper rim rays meet r_8 at a height of $+0.452$. These limiting rays are marked V in Fig. 136, their paths being shown in Fig. 137. If there were no vignetting the upper and lower rays would be limited only by the diaphragm, these rays being marked S in Fig. 136. It is clear that the vignetting has proved to be very beneficial, especially for the lower rays at 22°; these would cause a bad one-sided haze due to higher order coma if they were not vignetted out in this way.

A glance at this graph reveals that there is a small residual of negative coma, requiring a small negative bending of the front and rear crowns to remove it. A Δc of -0.005 was found to be sufficient to make all three curves quite straight. To gild the lily, trifling bendings were applied to remove small residuals of LA' and X'_t, namely, $\Delta c_1 = -0.0005$ and $\Delta c_3 = -0.0025$. The final lens was then scaled to a focal length of 10, with the following specification:

c	d	n_D
0.34138		
	0.369	1.6109
-0.13373		
	0.252	
-0.25288		
	0.111	1.6053
0.18937		
	0.221	
	0.221	
-0.18937		
	0.111	1.6053
0.25288		
	0.252	
0.13357		
	0.369	1.6109
-0.36091		

with $f' = 10.0$, $l' = 9.1734$, Petzval sum $(10) = 0.0342$, LA' $(f/6) = 0.0143$, LZA $(f/8.5) = -0.0193$; for $22°$: $X'_s = -0.0155$, $X'_t = 0$, distortion $= 0.218\%$, lateral color $= -0.0004$; zonal chromatic aberration $= 0.0001$. The aberrations of this final system are shown in Fig. 137. One interesting point here is that after all the changes in power and bending that have been made, radii (2) and (7) are almost identical. It might be a significant manufacturing economy to make them identical by a further trifling bending to one or both of the positive elements.

Lenses of this dialyte type perform admirably and can be designed with apertures up to about $f/3.5$; the field, however, is limited to about 22 or 24° from the axis.

III. A DOUBLE GAUSS TYPE LENS

The mathematician Gauss once suggested that a telescope objective could be made with two meniscus-shaped elements, the advantage being that

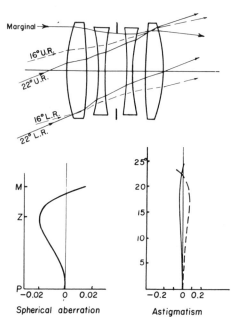

FIG. 137. An $f/6$ dialyte objective.

such a system would be free from spherochromatism. However, this arrangement has other serious disadvantages and it has not been used in any large telescope. Alvan G. Clark tried to use it with no success, but with considerable insight he recognized that two such objectives mounted symmetrically about a central stop might make a good photographic lens. He patented[4] the idea in 1888, and a lens of this type called the Alvan G. Clark lens was offered for sale by Bausch and Lomb from 1890 to 1898. The same type was also used in the Ross Homocentric, the Busch Omnar, and the Meyer Aristostigmat. An unsymmetrical version was later used by Kodak in their Wide Field Ektar lenses.

The design is suitable for a low-aperture wide-angle objective, the design procedure following closely the design of the dialyte just described. However, the glasses must be much further apart on the V-n chart than before, possible types being

(a) Dense flint: $n_D = 1.6170$, $V = 36.60$, $n_F = 1.62904$, $n_C = 1.61218$
(b) Dense barium crown: $n_D = 1.6109$, $V = 57.20$, $n_F = 1.61843$, $n_C = 1.60775$

---

[4] A. G. Clark, U.S. Patent 399,499, filed Oct. 1888.

Using these glasses, the formulas given on p. 237 can be solved for the two lens powers in the rear component, assuming zero L'_{ch} and a smaller Petzval sum such as 0.028 on a focal length of 10. The powers are much smaller than before and the air space much larger:

$$\phi_a = -0.2937, \qquad c_a = -0.4760$$

$$\phi_b = 0.3376, \qquad c_b = 0.5526$$

$$d = 0.5657$$

Since the lens elements are to be meniscus in shape, we can start by selecting bendings having $c_1 = 1.9c_a$ and $c_3 = -0.17c_b$. For a half-system of aperture $f/16$ we could try thicknesses of 0.1 for the flint element having a diameter of 0.9, and 0.3 for the crown element of diameter 1.9. (This is actually thicker than necessary, and 0.23 would have been better.) As before, after inserting the thicknesses we scale each element back to its original power, and we calculate the air space required to restore the separation between the adjacent principal planes. The stop is placed conveniently at 0.15 in front of the vertex of the negative element.

Having assembled the double lens, we find that its focal length is 6.255, the zonal chromatic aberration is -0.00398, and the Petzval sum for a focal length of 10 is 0.0249. As before, we proceed to correct the chromatic aberration and Petzval sum by changing the outer spaces and the powers of the positive elements, maintaining symmetry at all times. The double graph for these changes is shown in Fig. 138.

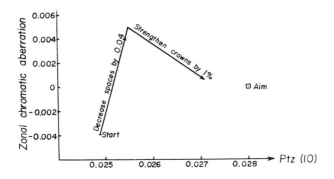

Fig. 138. Double graph for chromatic aberration and Petzval sum.

The graph suggests that we should strengthen both crown elements by 1.47% and decrease the outer spaces by 0.0466. These changes give the following front half-system (the rear is identical):

c	d	n_D
0.6491		
	0.3	1.6109
0.0952		
	0.1860	
0.4141		
	0.10	1.6170
0.9044		
	0.15	

with $f' = 6.0810$, aperture $= f/8$, zonal chromatic $= 0.00015$, Petzval sum $(10) = 0.0279$. We regard these residuals as acceptable and proceed to correct the spherical aberration and tangential field curvature by bending the elements in a symmetrical manner. The aberrations of this system were found to be

$$f/8 \text{ spherical aberration} = -0.0652$$

$$f/11.3 \text{ zonal aberration} = -0.0311$$

$$LA' + LZA = -0.0963$$

$$(32°)X'_s = 0.1538, \qquad X'_t = 0.0937$$

The results of separately bending the crowns and flints are shown in the double graph in Fig. 139, and a few trials indicate that we should bend the

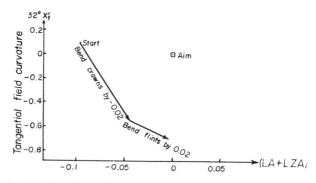

FIG. 139. Double graph for spherical aberration and field curvature.

front crown by 0.0128 and the front flint by 0.0627 to remove both the spherical aberration and tangential field curvature simultaneously. These changes give focal length $= 5.8951$ $LA' = -0.0020$, $LZA = -0.0003$

$$X'_s = 0.1196, \qquad X'_t = -0.0049$$

which are acceptable, but now we find that the bendings have upset our previous corrections for Petzval and chromatic aberration, which have become $Ptz(10) = 0.0271$, and zonal chromatic $= -0.0067$. It is characteristic of meniscus elements that any change in the lens shape affects all aberrations, an unfortunate property that makes the design of a Gauss-type lens much more difficult and time-consuming than the design of a comparable dialyte lens.

To remove these Petzval and chromatic residuals, we return to the graph in Fig. 138, which suggests that we should make a further reduction in the air spaces of 0.037, and strengthen the crowns by 0.191%. These changes in their turn upset the spherical and field corrections, so we have to make further small bendings for these, and so on back and forth until all four aberrations are correct. The system then is as follows:

c	d	n_D
0.6733		
	0.3	1.6109
0.1183		
	0.145	air
0.4768		
	0.1	1.6170
0.9671		
	0.15	
	(symmetrical)	

with focal length $= 5.9394$, $Ptz(10) = 0.0278$; for $32°$: $X'_s = 0.0674$, $X'_t = -0.0134$; $LA' = -0.0015$, $LZA = 0.0000$, zonal chromatic aberration $= -0.0008$.

Finally, we come to the correction of distortion and lateral color:

$$32° \text{ distortion} = 1.28\%, \qquad 32° \text{ lateral color} = 0.0014$$

These were reduced by shifting 3% of the power of the front crown element to the rear crown, giving 0.70% distortion and -0.0001 of lateral color. However, since this move upset everything, it was necessary to return to the previous graphs and repeat the whole design process once or twice more. After scaling up to a focal length of 10.0, the final system is as follows:

c	d	n
0.38600		
	0.5083	1.6109
0.06787		
	0.2355	
0.28732		
	0.1694	1.6170
0.57670		
	0.2542	
	0.2542	
−0.57670		
	0.1694	1.6170
−0.28732		
	0.2355	
−0.07201		
	0.5083	1.6109
−0.40990		

with $f' = 10.0$, $l' = 8.9971$, $\text{Ptz}(10) = 0.00279$, zonal chromatic aberration $= 0.00030$, $LA'\ (f/8) = 0.00046$, $LZA\ (f/11) = 0.00225$, stop diameter $(f/8) = 0.5149$. Then

Field (deg)	X'_s	X'_t	Distortion (%)	Lateral color
32	0.0617	−0.1303	0.660	−0.00034
25	−0.0529	0.0586	0.266	−0.00086
15	−0.0456	0.0239	0.065	−0.00064

A scale drawing of this lens, together with its aberration graphs, is shown in Fig. 140. It will be seen that the zonal aberration is of the unusual over-corrected type and that the crown elements are quite unnecessarily thick. The accompanying meridional ray plots in Fig. 141 indicate that coma is negligible, but there is a considerable degree of overcorrected oblique spherical aberration, which increases as the obliquity is increased. This is typical of all meniscus lenses.

To complete the work we must determine how much vignetting should be introduced, mainly to cut off the ends of the curves in Fig. 141. Our procedure will be to decide to accept a maximum departure of the graphs from the principal ray by, say, ± 0.025, and cut off everything beyond that limit. The vignetted rays have been drawn in the lens diagram, Fig. 140. The limiting rays are marked V on the ray plots, the extreme unvignetted rays

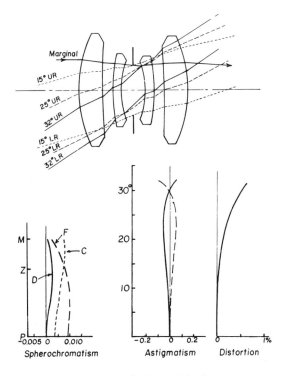

FIG. 140. An $f/8$ Gauss objective.

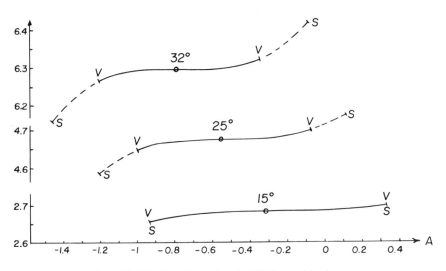

FIG. 141. Meridional ray plots for $f/8$ Gauss objective.

that just fill the diaphragm being marked S. It will be seen that the 15° beam is unvignetted. In view of this, the following limiting surface apertures are recommended for this lens:

Surface	Clear aperture
1	1.061
2	0.890
3	0.603
4	0.520
Stop	0.515
5	0.510
6	0.596
7	0.859
8	1.023

This completes the design.

If a higher aperture than $f/8$ is desired, it is necessary to thicken the negative elements considerably, and introduce achromatizing surfaces into them. The process for the design of the $f/2$ "Opic" lens of this type has been described by H. W. Lee.[5]

---

[5] H. W. Lee, The Taylor–Hobson $F/2$ anastigmat, *Trans. Opt. Soc.* **25**, 240 (1924).

CHAPTER 13

Unsymmetrical Photographic Objectives

I. THE PETZVAL PORTRAIT LENS

This ancient lens was the first photographic objective to be deliberately designed rather than being put together by an empirical selection of lenses out of a box. It consists of two fairly thin achromats spaced widely apart with a central stop.[1] It has excellent correction for spherical aberration and coma, but because the Petzval sum is uncorrected, the angular field is limited by astigmatism to about 12 to 15° from the axis. Modified forms of the Petzval lens are still used, mainly for the projection of 16 and 8 mm movie films, although if a negative field flattener is added close to the image plane the lens becomes a true anastigmat, and in this form it has been used as a long-focus lens for aerial reconnaissance purposes.

The front component of the original Petzval design of 1839 was an ordinary $f/5$ telescope doublet. It is possible that Petzval attempted to assemble two identical lenses symmetrically about a central stop, in order to raise the aperture to $f/3.5$ for use with the slow daguerreotype plates of the time, but the aberrations were so bad that he had to separate the two elements in the rear component and bend them independently to correct the spherical aberration and coma. Later, in 1860, J. H. Dallmeyer turned[2] the rear component around, with the crown element leading, and he thus obtained a lens that was better than the Petzval design near the middle of the field, but the inevitable uncorrected astigmatism was so great that the two designs are virtually indistinguishable. In 1878 F. von Voigtländer[3] found

---

[1] M. von Rohr, "Theorie und Geschichte des photographischen Objektivs," p. 250. Springer, Berlin, 1899.

[2] J. H. Dallmeyer, Brit. Patent 2502/1866; U.S. Patent 65,729.

[3] F. von Voigtländer, Brit. Patent 4756/1878.

that by suitably bending the front component of the Dallmeyer type he could cement the rear component also, and it is this last arrangement that is used today as a small projection lens of high aperture.

A. THE PETZVAL DESIGN

In designing a Petzval portrait lens it is customary to make both doublets of the same diameter and to mount the stop approximately midway between them. If the front doublet consists of the familiar form with an equiconvex crown, this stop position has the effect of making the tangential field of the front component somewhat backward-curving, and to correct this requires a positive rear component somewhat weaker than the front component. To correct the spherical aberration as well as the OSC and to flatten the tangential field, we find that we must select glass types having a rather large V difference; with the refractive indices used by Petzval, 1.51 and 1.57, a V difference of at least 18 is required. In the present examples the following Schott glasses are used:

(a) Crown: K-1, $n_e = 1.51173$, $n_F - n_C = 0.00824$, $V_e = 62.10$
(b) Flint: LF-6, $n_e = 1.57046$, $n_F - n_C = 0.01325$, $V_e = 43.05$

The V difference is 19.05.

1. The Front Component

For the front component we adopt a thin-lens focal length of 10 and a clear aperture of 1.8; this aperture may have to be adjusted later after the actual focal length of the system has been determined. For this front lens the thin-lens formulas give $c_a = 0.63706$ and $c_b = -0.30618$. Assuming an equiconvex crown, our front component is as follows:

c	d	n
0.31853		
	0.4	1.51173
−0.31853		
	0.12	1.57046
$(D - d)$ 0.086680		

Assuming an air space of 2.6, the $10°$ principal ray enters at $L_{pr} = 2.054$ and crosses the axis midway between the two lenses.

2. *The Petzval Rear Component*

For a Petzval-type rear component, we may start with the following arbitrary setup:

	c	d	n
	0.25		
		0.12	1.57046
	0.6		
		0.025516	
	0.55		
		0.4	1.51173
$(D - d)$	−0.017292		

with $f' = 6.1898$, $l' = 3.9286$, LA' $(f/3.44) = 0.0005$, OSC $(f/3.44) =$ 0.001944. The focal length and aberration data given here are calculated for the complete system. The space between the two rear elements was determined so that they would be in edge contact at a diameter of 1.8. As the design proceeds this separation must be recalculated at each setup to maintain the edge–contact condition.

The best way to correct the spherical aberration and coma is to bend the two rear elements separately and plot a double graph, Fig. 142. The graph

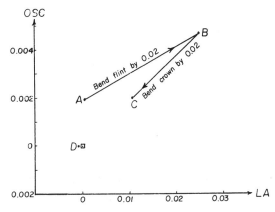

FIG. 142. Double graph for rear component of Petzval Portrait lens $(f' = 6.2)$.

data are:

(a) Original setup: $LA' = 0.000449$, $OSC = 0.001944$
(b) Bend flint by 0.02: $LA' = 0.024885$, $OSC = 0.004688$
(c) From *B*, bend crown by 0.02: $LA' = 0.010455$, $OSC = 0.001965$

Extrapolating in the usual way, and because the graphs are remarkably straight, we quickly reach the aplanatic form (setup D):

c	d	n
0.27		
	0.12	1.57046
0.62		
	0.018802	
0.5841		
	0.40	1.51173
$(D-d)$ 0.0220382		

with $f' = 6.2206$, $l' = 3.9233$, LA' $(f/3.46) = -0.0009$, OSC $(f/3.46) = -0.00003$. The fields along the computed $10°$ principal ray were $X'_s = -0.0597$, $X'_t = -0.0123$.

To move the fields backward, we must weaken the entire rear component. A few trials indicate that c_c should be reduced by 0.025, and after recorrecting the spherical and chromatic aberrations and the OSC we obtain the following solution (E):

c	d	n
0.27		
	0.12	1.57046
0.595		
	0.023158	
0.5495		
	0.40	1.51173
$(D-d)$ 0.0287696		

with $f' = 6.4012$, $l' = 4.0408$, LA' $(f/3.56) = 0.0030$, LZA $(f/5) = -0.0021$, OSC $(f/3.56) = -0.00002$, Ptz $(10) = 0.0811$. Then

Field (deg)	X'_s	X'_t	Distortion (%)
15	−0.1034	0.1551	0.32
10	−0.0571	0.0007	0.11

These aberrations are plotted in Fig. 143.

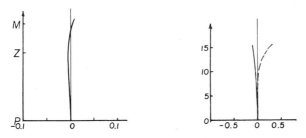

FIG. 143. Aberrations of setup E ($f' = 6.4$).

The final check on our system is made by drawing a meridional ray plot at 10° obliquity, Fig. 144a. The abscissas are the height of each ray at the

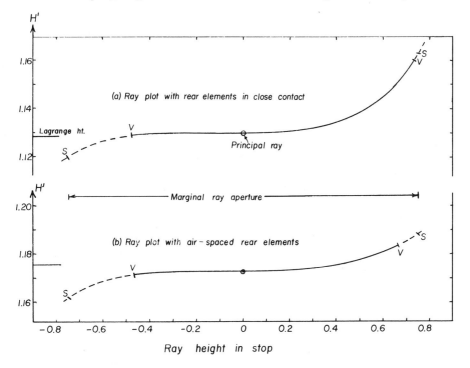

FIG. 144. Ray plots for Petzval objectives at 10°.

stop, the height of the marginal ray at the stop being shown on the graph. However, because of vignetting at the front and rear surfaces, which are assumed to have a free aperture of 1.8, only a part of the graph is valid. The upper and lower vignetted rays are indicated by VV, whereas the limiting rays through the top and bottom of the stop are marked SS on this graph. It should be noted particularly that the middle of the curve is straight and level as a result of the good correction of OSC and the flat tangential field at $10°$, but the upper end of the curve rises precipitously because of the extremely high values of the angles of incidence in the rear air space. Furthermore, the tangential field at $15°$ becomes rapidly more backward-curving for the same reason.

The best way to improve both these conditions is to increase the air space between the two rear elements. We will try a fixed air space of 0.15 and repeat the entire design. Following the same procedure, and plotting the usual graphs, gives us this final solution for the rear component, using the same front component and central air space as before:

c	d	n
0.25		
	0.12	1.57046
0.54		
	0.15	
0.468		
	0.40	1.51173
$(D-d)$ 0.0028107		

with $f' = 6.6685$, $l' = 4.2468$, LA' $(f/3.70) = 0.0012$, LZA $(f/5.2) = -0.0031$, OSC $(f/3.70) = 0.00001$, Ptz $(10) = 0.0804$. Then

Field (deg)	X'_s	X'_t	Distortion (%)	Lateral color
15	-0.1105	0.0157	-0.95	
10	-0.0553	-0.0002	-0.28	-0.00049

The $10°$ meridional ray plot for this lens, to the same scale as before, is shown in Fig. 144b. It will be seen that this design is much better than the previous one, and indeed almost all of the Petzval portrait lenses made since 1840 have had a wide space between the two rear elements. A sectional drawing of this system, with its aberration graphs, is shown in Fig. 145.

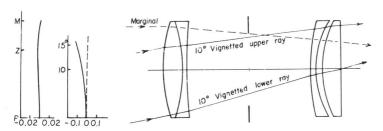

FIG. 145. The final Petzval-type lens.

B. The Dallmeyer Design

To design a lens of the Dallmeyer type, we can start by merely turning around the rear component of the last system, recomputing the last radius by the $D - d$ method, and tracing enough rays to evaluate the system. The crown element was made slightly thinner as it appeared to be too thick before. It was found that the spherical aberration had become decidedly undercorrected and the *OSC* overcorrected, and so a double graph was plotted by which these aberrations could be corrected, using suitable bendings of both rear elements. This graph is shown in Fig. 146, and it led us to

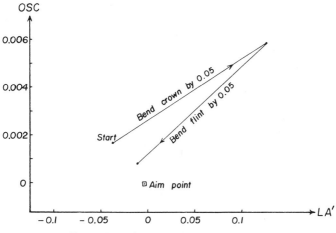

FIG. 146. Double graph for Dallmeyer lens.

the following rear system:

	c	d	n
	0.0722		
		0.35	1.51173
	−0.3930		
		0.15	
	−0.4600		
		0.12	1.57046
$(D - d)$	−0.1408571		

with $f' = 6.9991$, $l' = 4.3100$, LA' ($f/3.89$) $= -0.0125$, $OSC = -0.00006$; for $10°$: $X'_s = -0.0703$, $X'_t = -0.0423$.

In an attempt to correct the inward-curving tangential field, the third element was weakened by 0.1, and since this had the effect of reducing the relative aperture of the system, the front clear aperture was increased at the same time from 1.8 to 2.0. This required a recomputation of c_3 for achromatism by the $D - d$ method. It was found that the same double graph could be used, and a few trials gave the following rear component:

	c	d	n
	−0.0741		
		0.35	1.51173
	−0.4393		
		0.15	
	−0.5283		
		0.12	1.57046
$(D - d)$	−0.2880611		

with $f' = 7.3340$, $l' = 4.6524$, LA' ($f/3.67$) $= 0.0018$, $OSC =$ zero; for $10°$: $X'_s = -0.0481$, $X'_t = 0.0358$. Obviously we have gone too far in our weakening of the rear component, so we decided to strike a compromise and repeat the design. The final complete system then became as follows:

	c	d	n
	0.31853		
		0.40	1.51173
	−0.31853		
		0.12	1.57046
$(D - d)$	0.0847414		

Cont'd

Cont'd

	c	d	n
		2.6	
	0		
		0.35	1.51173
	−0.411		
		0.15	
	−0.4884		
		0.12	1.57046
(D − d)	−0.2114208		

with $f' = 7.1831$, $l' = 4.4796$, LA' $(f/3.59) = -0.0014$, LZA $(f/5.1) = -0.0136$, OSC $(f/3.59) = -0.00007$, Petz (10) = 0.0774. Then

Field (deg)	X'_s	X'_t	Distortion (%)	Lateral color
15	−0.1278	0.0359	1.54	
10	−0.0599	−0.0049	0.18	0.000692

A section of the lens is shown in Fig. 147 along with the graphs of the

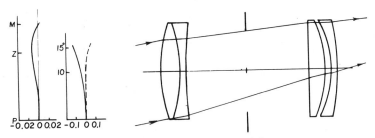

FIG. 147. A Dallmeyer-type portrait lens.

aberrations. A meridional ray plot is shown in Fig. 148, where it will be seen to be somewhat flatter than the better of the two Petzval designs. However, the large astigmatism would swamp this slight improvement. The zonal spherical aberration, although still small, is about four times as great as for the Petzval form.

II. THE DESIGN OF A TELEPHOTO LENS

A telephoto lens is one in which the " total length " from front vertex to focal plane is less than the focal length; telephoto lenses are used wherever the length of the lens is a serious consideration.

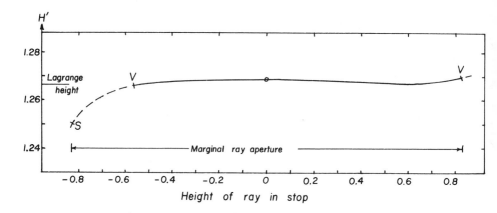

FIG. 148. Ray plot for Dallmeyer portrait lens at 10°.

Most telephoto objectives contain a positive achromat in front and a negative achromat behind, the lens powers being calculable when the focal length F, the total length kF, and the lens separation d are all given

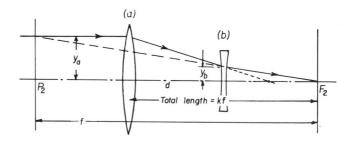

FIG. 149. Thin-lens layout of a telephoto system with a distant object.

(Fig. 149). The factor k is known as the telephoto ratio, and its value is ordinarily about 0.8.

In terms of thin lenses, the ratio

$$\frac{y_b}{y_a} = \frac{f_a - d}{f_a} = \frac{kF - d}{F}$$

whence

$$f_a = \frac{Fd}{F(1-k)+d}$$

For lens (b), we have $l = f_a - d$ and $l' = kF - d$. Therefore,

$$\frac{1}{f_b} = \frac{1}{kF - d} - \frac{1}{f_a - d}$$

whence

$$f_b = \frac{(f_a - d)(kF - d)}{f_a - kF}$$

As an example, if $F = 10.0$ and $k = 0.8$, we can plot graphs of the focal lengths of the two components against the lens separation d (Fig. 150). It is

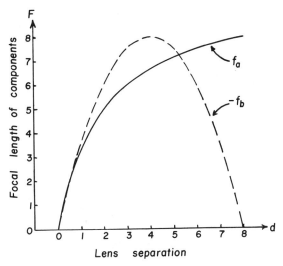

FIG. 150. Relation between lens powers and separation.

clear that as the separation is increased, both the front and rear lenses become weaker. Indeed, the power of the rear negative component reaches its minimum value when that lens lies midway between the front positive component and the focal plane. However, as the separation is increased the diameters of the lenses must also be increased to reduce the vignetting.

For our present design we will assume a thin-lens separation d equal to 3.0. This will require a positive front component with $f_a = 6.0$ and a negative rear component with focal length $f_b = -7.5$. Since the two chromatic aberrations will be controlled by a suitable choice of glass dispersions at the end, we adopt refractive indices such that a range of dispersions is available. The crown index is therefore set at 1.524, for which there are glasses with V values ranging from about 51 to 65, and the flint at 1.614, for which V-values exist between about 37 and 61.

For a start, let us suppose that the chosen glasses are K-3 and F-3:

(a) K-3: $n_e = 1.52031$, $\Delta n = n_F - n_C = 0.00879$, $V_e = 59.19$
(b) F-3: $n_e = 1.61685$, $\Delta n = n_F - n_C = 0.01659$, $V_e = 37.18$

with the V difference $= 22.01$. For the front component then, $c_a = 0.8615$, while for the rear component $c_c = -0.6892$. We may assume an equiconvex crown for the front and an equiconcave crown in the rear. We assign suitable thicknesses for a clear aperture of 1.8 (i.e., an aperture of $f/5.6$), and we consider an angular semifield of $10°$. For every setup we calculate the last radii of the front and rear components to yield the desired focal lengths of $+6.0$ and -7.5, respectively, and we determine the central air space so that the separation of adjacent principal points is 3.0. The stop is assumed to be in the middle of the air space. Our starting system is as follows:

	c	d	n
	0.4308		
		0.50	1.524
	-0.4308		
		0.15	1.614
(for f')	0.04155		
		2.517648	
	-0.3446		
		0.15	1.524
	0.3446		
		0.50	1.614
(for f')	-0.023990		

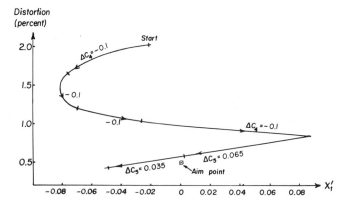

FIG. 151. Double graph for distortion and field curvature.

with $f' = 10.0$, $l' = 4.5205$, LA' $(f/5.6) = 0.3022$, OSC $(f/5.6) = -0.0260$; for $10°$: distortion = 2.04%, $X_t' = -0.0218$. We now proceed to change the rear component to correct distortion and tangential field curvature, using c_4 and c_5 on a double graph, of course maintaining the thin-lens telephoto conditions by solving for c_6 and the central air space d_3' at all times. It is found that the graph for changes in c_4 bends back on itself but the graph for c_5 is quite straight (Fig. 151). The aim point at distortion = 0.5 and $X_t' = 0$ is nearly reached by the following setup:

	c	d	n
	0.4308		
		0.50	1.524
(unchanged)	−0.4308		
		0.15	1.614
	0.04155		
		3.058468	
	−0.7446		
		0.15	1.524
	0.4100		
		0.50	1.614
	−0.310175		

with $f' = 10$, $l' = 3.8945$, LA' $(f/5.6) = 0.6093$, OSC $(f/5.6) = -0.0161$; for $10°$: distortion = 0.580%, $X_t' = 0.0022$. These changes have led to considerable overcorrection of the spherical aberration, while the OSC is slightly smaller.

We now move to the front and plot a double graph of spherical aberration and OSC for changes in c_1 and c_2 (Fig. 152). The closest setup to the

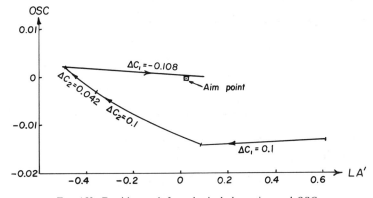

FIG. 152. Double graph for spherical aberration and OSC.

aim point at $LA' = 0.02$ and $OSC = 0$ is as follows:

	c	d	n
	0.4228		
		0.50	1.524
	−0.2888		
		0.15	1.614
	0.0551473		
		3.062479	
	−0.7446		
		0.15	1.524
(unchanged)	0.4100		
		0.50	1.614
	−0.310175		

with $f' = 10$, $l' = 3.8945$, LA' $(f/5.6) = 0.1017$, OSC $(f/5.6) = -0.00003$; for $10°$: distortion $= 0.878\%$, $X'_t = -0.3247$.

It is clear by now that in this type of lens every change affects every aberration, and it is not very profitable to go back and forth between the two double graphs. Instead, therefore, we will resort to the solution of four simultaneous linear equations in four unknowns, each equation being of the type

$$\Delta \text{ ab} = \sum (\partial \text{ ab}/\partial \text{ var}) \, \Delta \text{ var}$$

where ab signifies an aberration and var signifies a variable lens parameter (see Chapter 16, Section I).

The 16 coefficients of the type $(\partial \text{ ab}/\partial \text{ var})$ are found by trial, by applying a small change, say 0.1, to each variable in turn and finding its effect on each of the four aberrations. The coefficients were found to be:

Aberration	c_1	c_2	c_3	c_4
LA'	−5.488	−3.180	−0.6211	0.1416
OSC	0.01886	0.11995	−0.03421	−0.00590
Distortion (%)	−3.704	2.357	2.030	−4.021
X'_t	2.807	−2.439	−1.121	−1.274

These four simultaneous equations were solved to give the desired changes in the four aberrations, namely,

$$\Delta LA' = -0.08 \quad \text{(to yield } +0.02)$$

$$\Delta OSC = 0 \quad \text{(correct as is)}$$

$$\Delta \text{ distortion} = -0.38 \quad \text{(to give } +0.5)$$

$$\Delta X'_t = +0.32 \quad \text{(for zero)}$$

The solution of the equations was

$$\Delta c_1 = 0.0438, \qquad \Delta c_2 = -0.0333,$$

$$\Delta c_4 = -0.0906, \qquad \Delta c_5 = -0.0111$$

Applying these changes to our lens, and solving as before for the two focal lengths and the thin-lens separation, we get the following:

c	d	n
0.4666		
	0.50	1.524
−0.3221		
	0.15	1.614
(f) 0.092643		
	3.12194	
−0.8352		
	0.15	1.524
0.3989		
	0.50	1.614
(f) −0.372419		

with $f' = 10$, $l' = 3.74711$, LA' $(f/5.6) = 0.0248$, OSC $(f/5.6) = 0.00026$; for $10°$: distortion $= 0.534\%$, $X'_t = 0.0307$. These aberrations are almost correct, but a second solution using the same coefficients gave this final system:

c	d	n
0.4664		
	0.50	1.524
−0.3208		
	0.15	1.614
(f) 0.0926424		
	3.11078	
−0.8273		
	0.15	1.524
0.4083		
	0.50	1.614
(f) −0.3660454		

with $f' = 10.0$, $l' = 3.7618$, LA' $(f/5.6) = 0.0211$, LZA $(f/8) = -0.0108$,

OSC $(f/5.6) = -0.00001$; for 10°: distortion $= 0.50\%$, $X'_t = -0.0012$, $X'_s = 0.0261$.

We next trace a number of oblique rays at 10° obliquity and draw a meridional ray plot to determine the best stop position (Fig. 153). The

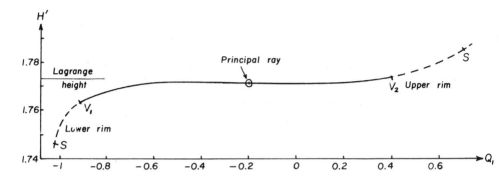

FIG. 153. Meridional ray plot of telephoto lens at 10° obliquity.

abscissas of this plot are conveniently the Q of each ray at the front surface. Since the lower end of this graph sags downward excessively, we must move the stop closer to the front than the midway position previously assumed. This puts the diaphragm at a distance of 0.5 from surface 3, and the limiting upper and lower rays that just fill the stop are shown by SS. When we set the front surface aperture at the diameter of the entering $f/5.6$ axial beam, namely, 1.786, the lower limiting ray is that shown at V_1. However, the graph indicates that we can safely increase the diameter of the rear aperture to 1.94, so that the upper limiting ray is that shown at V_2. The graph also indicates the presence of a small amount of overcorrected oblique spherical aberration, which is normal in lenses of this type. The principal ray now has a starting Q_1 of -0.2 for the 10° beam, or $L_{pr} = 1.1518$. Keeping this L_{pr} value we can add principal rays at 7° and 12°, giving the following:

Field (deg)	X'_s	X'_t	Distortion (%)
12	0.0018	-0.1611	0.26
10	0.0322	-0.0241	0.47
7	0.0189	0.0193	0.35

These results are plotted in Fig. 154.

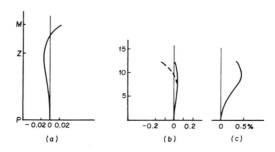

FIG. 154. Aberrations of a telephoto lens. (a) Spherical, (b) astigmatism, (c) distortion.

The lens in its present configuration is shown in Fig. 155. The telephoto

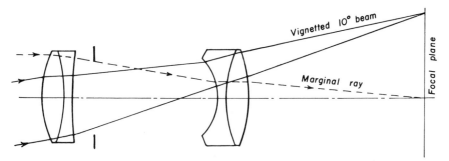

FIG. 155. Final telephoto design showing $f/5.6$ marginal ray and limiting rays at $10°$.

ratio is 81.7%, and the lens could be used in a focal length of 120 mm or more on a 35 mm camera. The aperture could be slightly increased, especially if the angular field were less than $10°$.

To wind up the design we must select real glasses for achromatism. The $D - d$ values along the marginal ray in the four lens elements are given, together with the products $(D - d)\,\Delta n$ for a first glass selection as follows:

Lens element	a	b	c	d
$D - d$	-0.314961	0.156459	0.073780	-0.045341
Glass type	BK-8	F-3	KF-7	F-3
$n_F - n_C = \Delta n$	0.00818	0.01659	0.01021	0.01659
Product $(D - d)\,\Delta n$	-0.0025764	0.0025957	0.0007533	-0.0007522
		0.0000193		0.0000011

$\sum = 0.0000204$

Adjusting the catalog indices of these glasses for C and F light by the same amount as the n_e was in error and tracing $10°$ principal rays in C and F gave the lateral color as $H'_F - H'_C = -0.006525$, which was considered excessive. Since the lateral color takes the same sign as the longitudinal color of the rear component and the opposite sign of that of the front component, it is clear that the $\sum (D - d) \Delta n$ of the rear should be more positive, while that of the front component should be more negative.

A second glass selection was as follows:

Lens element	a	b	c	d	
Glass type	BaLK-3	F-3	K-4	BaF-5	
n_e	1.52040	1.61685	1.52110	1.61022	
V	60.58	37.19	57.58	49.49	
Δn	0.00859	0.01659	0.00905	0.01233	
$(D - d) \Delta n$	−0.0027055	0.0025957	0.0006677	−0.0005591	
	−0.0001098		0.0001086		$\sum = -0.0000012$

Now the lateral color was found to be $+0.00088$, which was much better. No further improvement was possible using Schott glasses, and so the design was considered complete. It is, of course, necessary to repeat the final stages with the actual refractive indices of the chosen glasses, the procedure being to trace a paraxial ray with the true indices, and to adjust the curvature of each surface to maintain the ray-slope angle after each surface at its former value. Any small aberration residuals that appear can be removed by solving the four simultaneous equations again, assuming that the former 16 rate-of-change coefficients are still valid.

There is, of course, no magic in the initial choice of refractive indices, and it is possible that a better design could be obtained by a different choice.

III. THE PROTAR LENS

In 1890 Paul Rudolph of Zeiss[4] had the idea of correcting the spherical aberration of a new-achromat landscape lens by adding a front component resembling the front component of a Rapid Rectilinear but with very little power. The thought was that the strong cemented interface in the front component could be used to correct the spherical aberration, and that it would have little effect on field curvature because the principal rays would

---

[4] M. von Rohr, "Theorie und Geschichte des photographischen Objektivs," p. 364. Springer, Berlin, 1899.

be almost perpendicular to it. The cemented interface in the rear component would be used to flatten the field as in the new-achromat landscape lens.

This leaves us with four other radii to be determined. The fourth and sixth radii can be used for Petzval sum and focal length, as in the design of a new achromat, leaving the first and third radii for coma and distortion correction. The two chromatic aberrations are controlled by the final selection of glass dispersions.

As an example, we will first select suitable refractive indices. For elements (a) and (d) we may assume that $n_e = 1.6135$. In the Schott catalog we find many glasses having n_e lying close to this figure, with values of $V_e = (n_e - 1)/(n_F - n_C)$ lying between 37.2 and 59.1. For elements (b) and (c) we choose similarly $n_e = 1.5146$ for which V_e values are available between 51.2 and 63.6. Suitable thicknesses are, respectively, 0.25 and 0.4 for the front component and 0.1 and 0.4 for the rear component; the center space is set at 0.4 with the diaphragm midway. We do not use diaphragm position as a degree of freedom since we have enough degrees of freedom already; it is, however, advisable to keep the center space small to reduce vignetting.

For a first trial we may choose $c_1 = 0.5$, $c_2 = 1.2$, and $c_5 = 0.5$. We solve c_3 to make the front component afocal, and we determine c_4 and c_6 by trial and error to make the focal length equal to 10 and the Petzval sum 0.025. The first setup is as follows:

Setup A

c	d	n
0.5		
	0.25	1.6135
1.2		
	0.4	1.5146
0.417646		
	0.2	
	0.2	
−0.626156		
	0.1	1.5146
0.5		
	0.4	1.6135
−0.572960		

with $f' = 10$, $l' = 9.8120$, Ptz = 0.025, trim diameter = 1.5. A scale drawing of this lens is shown in Fig. 157. Tracing an $f/8$ marginal ray from infinity and a principal ray passing through the center of the stop at a slope of $-20°$

gives these starting aberrations:

$$LA' \text{ at } f/8 = -0.09026, \quad \left.\begin{array}{l} X'_s = -0.0610 \\ X'_t = +0.0169 \end{array}\right\} \quad \text{at} \quad U_{pr} = -17.90°$$

Since the main function of radius r_2 is to control the spherical aberration and the main function of r_5 is to flatten the field, we next proceed to vary c_2 and c_5 in turn by 0.05 and plot a double graph by means of which the spherical aberration and the tangential field curvature can be corrected. We assume that the desired values of these aberrations are $LA' = +0.15$ and $X'_t = 0$. The double graph in Fig. 156 indicates that we should make the following changes from the original setup A:

1. $\Delta c_2 = 0.034$. But c_2 was 1.2, so therefore try new $c_2 = 1.234$.
2. $\Delta c_5 = 0.009$. But c_5 was 0.5, so therefore try new $c_5 = 0.509$.

These changes give setup B, the thicknesses and refractive indices remaining as before:

Setup B

c	d
0.5	
	0.25
1.234	
	0.4
0.410355	
	0.2
	0.2
−0.637781	
	0.1
0.509	
	0.4
−0.579493	

with $f' = 10$, Ptz = 0.025, LA' $(f/8) = 0.1361$, LZA $(f/11) = -0.0724$; for 17.91°: $X'_s = -0.0668$, $X'_t = -0.0023$.

Before making any further changes in LA' and X'_t, we must decide whether the aim point that we have chosen is the best. Certainly the zonal spherical aberration is about right, and so we will maintain our aim for spherical aberration at +0.15. However, to study the field requirements, it is necessary to trace several more principal rays at higher obliquities and plot the astigmatism curves. These rays give the following tabulation:

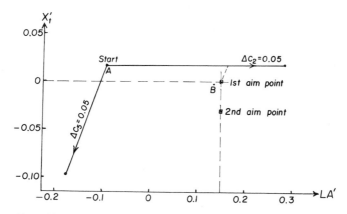

FIG. 156. Double graph for spherical aberration and field curvature.

Field angle at object (deg)	Field angle in stop (deg)	X'_s	X'_t	Distortion (%)
−35.00	−40	+0.17590	+0.24096	−2.52
−26.62	−30	−0.05159	+0.10503	−1.25
−17.91	−20	−0.06677	−0.00230	−0.51

A plot of these field curves (Fig. 157) indicates at once that a much better

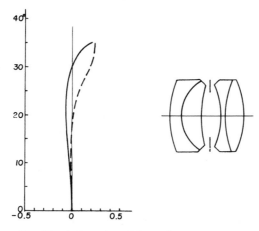

FIG. 157. Astigmatism of Protar lens, setup B.

aim point for the sag of the tangential field X'_t would be at -0.03, and this value will be used from now on.

We next proceed to correct the coma and distortion. The OSC of setup B is found to be -0.00399 at $f/8$, and both the coma and distortion are clearly excessive. Our free variables are now c_1 and the power of the front component. By using $y = 1$ for the paraxial ray, the power of the front component is given directly by u'_3, and any desired value of this angle can be obtained by solving for c_3. Assuming for a start that both OSC and distortion should be zero, we plot a second double graph (Fig. 158), changing c_1 and u'_3. This graph indicates that we should make the following changes in the system:

1. $\Delta c_1 = 0.1227$. But c_1 is 0.5, so therefore we should try $c_1 = 0.6227$.
2. $\Delta u'_3 = 0.0117$. But u'_3 is zero, so therefore we should try $u'_3 = 0.0117$.

With these changes our lens becomes setup C:

Setup C

c	d	n
0.6227		
	0.25	1.6135
1.234		
	0.4	1.5146
0.570553		
	0.2	
	0.2	
-0.473548		
	0.1	1.5146
0.509		
	0.4	1.6135
-0.453187		

with $f' = 10$, Ptz $= 0.025$, power of front $= +0.0117$, OSC $(f/8) = -0.000402$, distortion $(18°) = -0.017\%$.

We will assume for the present that zero is a good aim point for distortion, but we must investigate the coma further. To do this we trace a family of rays entering the lens at $-17.23°$, and plot a graph connecting the height of incidence of each ray at the stop against the height H' of the ray at the paraxial focal plane. This graph, Fig. 159, indicates the presence of some negative primary coma with an upturn at both ends of the curve due to positive higher-order coma. Since the ends of the curve will probably be cut off by vignetting, it might be better to aim at, say, $+0.002$ of OSC at $f/8$

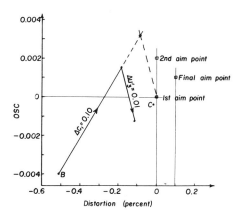

FIG. 158. Double graph for coma and distortion.

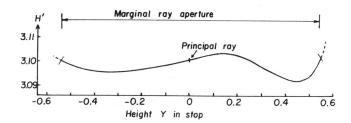

FIG. 159. Meridional ray plot for setup C (17°).

instead of zero. This will represent the aim point for OSC on future double graphs.

The spherical aberration of setup C is $+0.0908$ and the field curvature is given by $X_t' = +0.1531$. Reference to the first double graph, using the new aim point, indicates these changes:

1. $\Delta c_2 = +0.026$. But $c_2 = 1.234$, so therefore we should try $c_2 = 1.260$.
2. $\Delta c_5 = +0.0916$. But $c_5 = 0.509$, so therefore we should try $c_5 = 0.6006$.

These changes give us setup D:

Setup D

c
0.6227
1.260
0.564736
−0.466835
0.6006
−0.435009

with $f' = 10$, Ptz $= 0.025$, front power $= 0.0117$, LA' $(f/8) = 0.1422$; for $17.24°$: $X'_s = -0.0704$, $X'_t = -0.0331$.

The spherical aberration and field curvature of this system are acceptable. However, we find that the OSC at $f/8$ has now become -0.00313 and the $17°$ distortion -0.009%. Reference to the second double graph enables us to remove these residuals, and we then return to the first graph for spherical aberration and field curvature, and so on back and forth several times until all four aberrations are acceptable. The final setup is E:

Setup E

c	d	n_e
0.6445		
	0.25	1.6135
1.2466		
	0.4	1.5146
0.628369		
	0.2	
	0.2	
−0.383337		
	0.1	1.5146
0.5856		
	0.4	1.6135
−0.395628		

with $f' = 10$, Ptz $= 0.025$, power of front $= 0$, LA' $(f/8) = 0.1529$, LZA $(f/11) = -0.0487$, OSC $(f/8) = 0.00204$. The astigmatism and distortion values are as follows:

Field at object (deg)	Field in stop (deg)	X'_s	X'_t	Distortion (%)
− 33.63	− 40	+ 0.08847	− 0.27048	− 0.676
− 29.62	− 35	− 0.02411	− 0.07838	− 0.311
− 25.54	− 30	− 0.07381	− 0.03227	− 0.119
− 21.39	− 25	− 0.08298	− 0.03218	− 0.028
− 17.18	− 20	− 0.06921	− 0.03188	+ 0.007

These aberrations are plotted in Fig. 160, as well as the 17° meridional ray

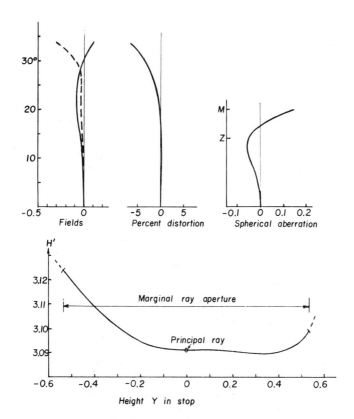

FIG. 160. Aberrations of final Protar design.

plot for the study of coma. It is clear from these graphs that we should have aimed at about +0.2% of distortion at 17° and about +0.001 of OSC at $f/8$. These values should be adopted in any future changes. The field and spherical aberration are just about right.

Our next task is to investigate the correction of the two chromatic aberrations by choice of glass. We first attempt to correct the $(D - d)\,\Delta n$ sum of each component separately, since this is the proper thin-lens solution to the problem. For this we use the true $\Delta n = n_F - n_C$ of each likely glass, ignoring the fact that the catalog refractive indices are not quite equal to those we have assumed so far. This gives the following:

Lens	Glass	n_e	Δn	$D - d$ for $f/8$ ray	$(D - d)\,\Delta n$	Sum
1	SK-8	1.61377	0.01095	0.108126	0.00118398 ⎫	+0.00003299
2	K-1	1.51173	0.00824	−0.139683	−0.00115099 ⎭	
3	KF-8	1.51354	0.01004	0.151056	0.00151660 ⎫	−0.00012649
4	SK-3	1.61128	0.01034	−0.158906	−0.00164309 ⎭	
					Total	−0.00009350

No suitable glasses were found by which we could have reduced the negative $(D - d)\,\Delta n$ sum in the rear component.

To calculate the lateral color at $17.18°$, we apply the same arithmetic error to the individual F and C indices as we have assumed for the e indices. This gives the following:

Lens	Glass	Nominal indices n_e	n_C	n_F	H'_C	H'_e	H'_F
1	SK-8	1.6135	1.60758	1.61853			
2	K-1	1.5146	1.51012	1.51836	3.090647	3.091227	3.091733
3	KF-8	1.5146	1.50920	1.51924			
4	SK-3	1.6135	1.60789	1.61823			

From these data we find that $H'_F - H'_C = +0.001086$. Now the lateral color in any lens takes the same sign as the longitudinal color of the rear component and the opposite sign to the longitudinal color of the front component. Hence to improve both the longitudinal and lateral color aberrations simultaneously we must make the $(D - d)\,\Delta n$ sum of the front component more positive. To do this we need a glass in lens (1) having a lower V number or a glass in lens (2) having a higher V number. Inspection of the chart enclosed with the glass catalog indicates that the only possible choice is to use BK-1 in place of K-1 in lens (2), because all other possible glasses have a refractive index differing too much from the e indices we assumed for the aberration calculations. This glass has $\Delta n = 0.00805$ giving a value of $\sum (D - d)\,\Delta n$ equal to $+0.00005953$ in the front component, or -0.00006696 for the whole system, and a lateral color of $H'_F - H'_C = +0.000790$. We must accept these residuals in the absence of other more extreme glass types.

Of course, the final stage is to repeat the design using the true n_e refractive indices, and then to adjust the clear apertures to give the desired degree of vignetting.

IV. DESIGN OF A TESSAR LENS

The Tessar[5] resembles the Protar in that the rear component is a new-achromat cemented doublet, but the front component is now an air spaced doublet rather than a cemented old-achromat. The cemented interface in the front component of the Protar was very strong, leading to a large zonal aberration, but the separated doublet in the Tessar gives so much less zonal aberration that an aperture of $f/4.5$ or higher is perfectly feasible. From another point of view, the Tessar can be regarded as a Triplet with a strong collective interface in the rear element; this interface has a threefold function: it reduces the zonal aberration, it reduces the overcorrected oblique spherical aberration, and it brings the sagittal and tangential field curves closer together at intermediate field angles.

1. Choice of Glass

It is customary to use dense barium crown for the first and fourth elements, a medium flint for the second, and a light flint for the third element. Possible starting values are therefore as follows:

Lens	Type	n_e	$\Delta n = n_F - n_C$	$V_e = (n_e - 1)/\Delta n$
a	SK-3	1.61128	0.01034	59.12
b	LF-1	1.57628	0.01343	42.91
c	KF-8	1.51354	0.01004	51.15
d	SK-3	1.61128	0.01034	59.12

2. Available Degrees of Freedom

Because of the importance of the cemented interface in the rear component, it is best to establish it at some particular value, say 0.45, and leave it there throughout the design. Since there is no symmetry to help us, we must correct every one of the seven aberrations, and also hold the focal length, by

---

[5] P. Rudolph, U.S. Patent 721,240, filed July 1902.

a suitable choice of the available degrees of freedom; this makes the design decidedly laborious especially if it is performed by hand on a pocket calculator.

In the front component we have the two powers, two bendings, and one air space. The second air space is held constant to reduce vignetting, while in the rear component we have only the two outer surface curvatures to be determined. We thus have seven degrees of freedom with which to correct six aberrations and hold the focal length. We must therefore use choice of glass to correct the seventh aberration.

Many possible ways of utilizing the various degrees of freedom could be tried. In this chapter we shall assign the available freedoms in the following way:

(a) The power of lens (a) and the dispersion of the glass in lens (d) will be used to control the two chromatic aberrations.

(b) The power of lens (b) will be solved to maintain the power of the front component at, say, -0.05 (a focal length of -20) for distortion correction.

(c) The curvature of the last surface, c_7, will be solved to make the overall focal length equal to 10.

(d) The front air space will in all cases be adjusted to make the Petzval sum equal to, say, 0.025.

(e) The spherical aberration will be corrected in all cases by a suitable choice of c_5.

(f) This leaves the bendings of lenses (a) and (b) to be used to correct the OSC and the tangential field curvature X'_t.

Our starting system A will be arbitrarily set as follows:

	c	d	n_e
	0.4		
		0.40	1.61128
	0		
		0.3518 (Ptz)	
	-0.2		
		0.18	1.57628
(u'_4)	0.406891		
		0.37	
		0.13	
	-0.05		
		0.18	1.51354
	0.45		
		0.62	1.61128
(u'_7)	-0.247928		

with $f' = 10$, Ptz $= 0.025$.

3. Chromatic Correction

Assuming that this is a reasonable starting system, we next trace an $f/4.5$ marginal ray in e light and find the $(D - d) \Delta n$ contribution of each lens element, as follows:

Element	(a)	(b)	(c)	(d)
$(D - d) \Delta n$	−0.00266942	0.00404225	0.00270365	−0.00387980
	0.00137283		−0.00117615	
	$\sum = 0.00019668$			

We could, of course, adjust the two components to make both totals separately zero, but it is then found that the lateral color $H'_F - H'_C$, calculated by tracing principal rays at $17°$, is strongly positive. Since lateral color takes the same sign as the longitudinal color of the rear component, we must have a considerable amount of negative $D - d$ sum in the rear and an equal positive sum in the front component. We will therefore try to increase the negative sum in the rear component by choosing a glass for element (d) with a higher dispersive power, that is, a lower V number. Such a glass is SK-8 with $n_e = 1.61377$, $\Delta n = n_F - n_C = 0.01095$, and $V_e = 56.05$. The slight alteration in refractive index requires a small adjustment of the system, giving setup B:

c	d	n_e
0.4		
	0.4	1.61128
0		
	0.3421	(air)
−0.2		
	0.18	1.57628
0.4051605		
	$\dfrac{0.37}{0.13}$	(air)
−0.05		
	0.18	1.51354
0.45		
	0.62	1.61377
−0.2444831		

with $f' = 10$, $l' = 8.851896$, Ptz = 0.025. An $f/4.5$ marginal ray gives $LA' = +0.30981$ and the following $D - d$ values:

Element	(a)	(b)	(c)	(d)
$(D - d) \Delta n$	-0.00266942	0.00405667	0.00272259	-0.00411584
		0.00138725		-0.00139325
		$\Sigma = -0.00000600$		

This sum is quite acceptable so far as longitudinal chromatic aberration is concerned. We must next check for lateral color. Tracing principal rays at $17°$ in F and C light tells us that the lateral color is $+0.000179$, which is also acceptable, so that now both chromatic aberrations are under control. Fortunately chromatic errors change so slowly with bendings that our future efforts at correcting spherical aberration, coma, and field curvature by bending the three components do not greatly affect the chromatic corrections.

4. Spherical Correction

Because elements (a) and (b) are working at about the minimum aberration positions, we cannot hope to correct spherical aberration by bending them. Thus we are obliged to control spherical aberration by bending the rear component, i.e., by changing c_5. This will be done by a series of trials at every setup from now on. We arbitrarily require LA' to be about $+0.098$; this will yield a zonal residual of about half that amount, giving excellent definition when the lens is stopped down, as it will almost always be in regular use. We then vary c_1 and c_3 to correct OSC and X'_t by means of a double graph (Fig. 161). The aim point will be at zero for both these aberrations.

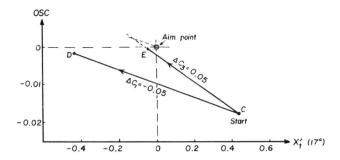

FIG. 161. Double graph showing effects of bending the first two elements of a Tessar. (In each case the Petzval sum and spherical aberration have been corrected first.)

Correcting the spherical aberration of system B by adjusting c_5 gives the following system C:

c	d	n_e
0.4		
	0.4	1.61128
0		
	0.2949	
−0.2		
	0.18	1.57628
0.3968783		
	0.37	
	0.13	
−0.080		
	0.18	1.51354
0.45		
	0.62	1.61377
−0.2632877		

with $f' = 10$, $l' = 9.035900$, Ptz $= 0.02499$, $LA' = 0.09724$, $OSC = -0.01778$. A principal ray traced at 20° through the stop emerges from the front of the lens at 17.3070°, with the following fields:

Angle: 17.31°, X'_s: 0.0716 X'_t: 0.4337,   distortion: 0.213%

The distortion is negligible, showing that our choice of $u'_4 = -0.05$ is about right, and we will continue with that value in what follows. The negative OSC is, however, much too large, and the tangential field is much too backward-curving.

5. *Correction of Coma and Field*

To plot a double graph, we make a trial change of $\Delta c_1 = -0.05$ from system C and restore everything to its original value (setup D). We then return to setup C and now change c_3 by 0.05, giving setup E. These changes are shown in Fig. 161. Following the usual procedure with a double graph, and making several small adjustments, we finally come up with setup F:

c	d	n_e
0.4126		
	0.40	1.61128
0.013442		
	0.2927	
−0.1366		
	0.18	1.57628
0.464462		
	0.37	
	0.13	
−0.0571		
	0.18	1.51354
0.45		
	0.62	1.61377
−0.247746		

with $f' = 10$, $l' = 8.9344$, $LA'(f/4.5) = 0.0958$, $LZA(f/6.4) = -0.0258$, OSC $(f/4.5) = 0$, $Ptz = 0.0250$. Then

Field angle (deg)	X'_s	X'_t	Distortion (%)
29.74	0.1607	0.1303	−1.42
25.61	0.0102	0.0871	−0.92
21.42	−0.0458	0.0305	−0.56
17.19	−0.0537	−0.0020	−0.32

The aberration graphs are shown plotted in Fig. 162. As a check on the

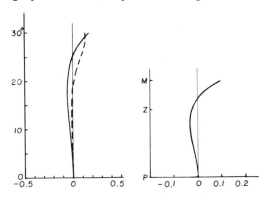

FIG. 162. Aberrations of system F.

coma we next trace a number of oblique rays entering parallel to the principal ray at 17.19° and raw a meridional ray plot (Fig. 163). It will be seen that

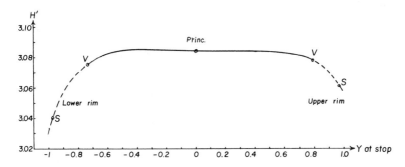

FIG. 163. Meridional ray plot of Tessar system F. Rays SS are through top and bottom of the stop. Rays VV represent vignetted limiting rays.

the two ends of this graph sag somewhat, but the middle part of the curve is straight. This is an indication of the presence of negative higher-order coma, and it cannot be usefully corrected by the deliberate introduction of positive OSC. A much better method of removing it is to introduce some vignetting. If we limit the clear aperture of each surface to the diameter of the entering $f/4.5$ axial beam, we shall cut off the ends of the ray plot in Fig. 163 to the marks VV shown, and we shall thus remove almost all of the higher-order coma without seriously reducing the image illumination. Figure 164 shows

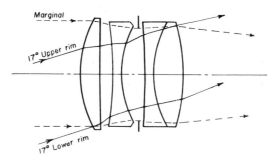

FIG. 164. Vignetting in system F, 17° beam.

the lens apertures so reduced and the path of the limiting oblique rays VV.

The astigmatic fields shown in Fig. 162 cross rather too high and the field is a little backward-curving. We shall therefore return to the double graph of Fig. 161 and establish a new aim point at $OSC = 0$ and $X'_t = -0.04$, which is by chance very close to setup E. After making several small adjustments in c_1 and c_3, and of course correcting the spherical aberration each time by c_5 and the Petzval sum by d'_2, we arrive at the following solution G:

c	d	n
0.4065		
	0.40	1.61128
0.0069273		
	0.3019	
−0.1421		
	0.18	1.57628
0.4596089		
	0.37	
	0.13	
−0.0579		
	0.18	1.51354
0.45		
	0.62	1.61377
−0.2486575		

with $f' = 10, l' = 8.925977$, Ptz $= 0.025$, LA' $(f/4.5) = 0.1029$, LZA $(f/6.4) = -0.0216$, OSC $(f/4.5) = 0$, $\sum (D - d) \, \Delta n = -0.00001096$, lateral color $H'_F - H'_C \, (17°) = -0.00031$. Then

Field (deg)	X'_s	X'_t	Distortion (%)
29.64	0.1224	−0.0283	−1.18
25.55	−0.0148	−0.0064	−0.77
21.38	−0.0619	−0.0257	−0.47
17.16	−0.0635	−0.0430	−0.27

The fields and aberration are shown plotted in Fig. 165.

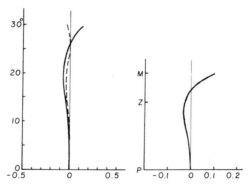

FIG. 165. Aberrations of Tessar design G.

We must now study the effect of changing the cemented interface c_6. This was arbitrarily set at 0.45, and we will next repeat the entire design with $c_6 = 0.325$. The resulting lens is decidely different from the previous design:

c	d	n
0.328		
	0.4	1.61128
−0.0757715		
	0.347	
−0.24		
	0.18	1.57628
0.3564288		
	0.37	
	0.13	
−0.135		
	0.18	1.51354
0.325		
	0.62	1.61377
−0.3216593		

with $f' = 10$, $l' = 9.20712$, LA' $(f/4.5) = 0.08714$, LZA $(f/6.4) = -0.03475$, OSC $(f/4.5) = 0$, $\sum (D - d) \Delta n = -0.0000707$, lateral color $(17°) = -0.00121$. Then

Field (deg)	X'_s	X'_t	Distortion (%)
25.41	0.0408	−0.0905	+0.12
21.38	−0.0244	0.0198	−0.04
17.22	−0.0413	0.0157	−0.06

These aberrations are shown in Fig. 166. The field is a little narrower than

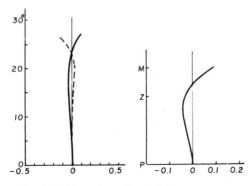

FIG. 166. Aberrations of second Tessar system.

before but quite satisfactory. It should be noted that both of the color aberrations are negative; to rectify this requires a small increase in the V number of the glass used for the rear crown element, say to SK-1, which has $n_e = 1.61282$ and $V_e = 56.74$, or SK-19 with $n_e = 1.61597$ and $V_e = 57.51$.

The chief matter requiring study is the meridional ray plot in Fig. 167,

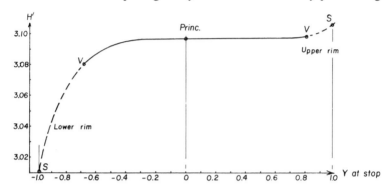

FIG. 167. Meridional ray plot for Tessar system with $c_6 = 0.325$ (17°).

which should be compared with the previous graph in Fig. 163. It is immediately clear that the change from $c_6 = 0.325$ to 0.45 has had the effect of raising the lower end of the curve and depressing the upper end. That is, strengthening c_6 has introduced some undercorrected oblique spherical aberration to the existing negative higher-order coma, with an improvement in the overall quality of the lens. The lower end of the curve needs cutting off more than the upper end, but obviously we cannot cut it back beyond the marginal ray aperture.

The best way to improve this Tessar is to raise the refractive indices, preferably above 1.6 in all elements. It is doubtful if changing the thicknesses would have any significant effect.

V. THE COOKE TRIPLET LENS

The English designer H. Dennis Taylor was led to this design[6] in 1893 by the simple consideration that if an objective were to consist of a positive lens and a negative lens of equal power and the same refractive index, the Petzval sum would be zero, and the system could be given any desired power by a suitable separation between the lenses. However, he quickly realized that the

---

[6] H. D. Taylor, Optical designing as an art, *Trans. Opt. Soc.* **24**, 143 (1923); also Brit. Patents 22607/1893, 15107/1895.

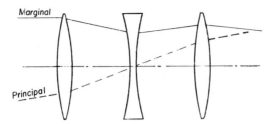

FIG. 168. The Cooke triplet lens.

extreme asymmetry of this arrangement would lead to an intolerable amount of lateral color and distortion, and so he split the positive element into two and mounted the negative element between them, thus making his famous triplet objective (Fig. 168). He also tried the alternative arrangement of dividing the negative element into two with the positive lens between, but this is much less favorable than the classic arrangement.

The triplet objective is tricky to design because a change in any surface affects every aberration, and the design would be impossibly difficult without a preliminary thin-lens predesign using Seidel aberrations. We assign definite required residuals for each primary aberration, and then by ray tracing determine the actual aberrations of the completed thick-lens system. If any aberration is excessive, we adopt a different value for that primary aberration and repeat the entire predesign. The thin-lens residuals that have been adopted here are the result of a series of such poor choices, the final thick system being satisfactory. Of course, in making a design differing from this in any important respect such as aperture, field, or glass selection, we would require a different set of Seidel aberration residuals, which would have to be found by trial.

A. THE THIN-LENS PREDESIGN OF THE POWERS AND SEPARATIONS

If we place the stop at the negative thin element inside the system, we can solve for the powers and separations of the three elements to yield specified values of the overall focal length and primary chromatic aberration, primary lateral color, Petzval sum, and one other condition that will eventually be used for distortion control. This last requirement might be the ratio of the two separations, the ratio of the powers of the outside elements, the ratio of the power of the combination of elements a and b to the power of the system, or some other similar criterion. We thus have five variables (three powers and two separations) with which to solve five conditions, after which we shall have three bendings to correct for the three remaining aberrations: spherical, coma, and astigmatism. Without this convenient division of the aberrations into two groups, those depending only on powers and separations and those depending also on bendings, the entire design process would be hopelessly complicated and almost impossible to accomplish.

The first part of the thin-lens predesign can be performed in several ways, the one employed here having been introduced by K. Schwarzschild about 1904. It uses the formulas for the contributions of a thin element to power, chromatic aberration, and Petzval sum, given on p. 207. These contributions may be written for each aberration in turn, as follows:

$$(y_a)\phi_a + (y_b)\phi_b + (y_c)\phi_c = (u'_0 - u_a) = y_a\Phi \quad \text{if} \quad u_a = 0 \quad \text{(power)}$$

$$(y_a^2/V_a)\phi_a + (y_b^2/V_b)\phi_b + (y_c^2/V_c)\phi_c = -L'_{\text{ch}}u_0'^2 \quad \text{(chromatic)}$$

$$(1/n_a)\phi_a + (1/n_b)\phi_b + (1/n_c)\phi_c = \text{Ptz} \quad \text{(Petzval)}$$

These three equations are linear in the three lens powers, and they can be easily solved for the powers once we know the three axial-ray heights y_a, y_b, and y_c. The first of these, y_a, is known when the focal length and F number are known, but y_b and y_c must be found by trial to satisfy the remaining two conditions, namely, the correction of lateral color and the ratio of the two separations $S_1/S_2 = K$. Reasonable starting values of the other ray heights are $y_b = 0.8y_a$ and $y_c = 0.9y_a$.

As an example, we will proceed to design an objective of focal length 10.0 and aperture $f/4.5$ covering a field of $\pm 20°$. We shall assume that $K = 1$, and use the following types of glass:

(a, c)   SK-16, $n_D = 1.62031$, $n_F - n_C = 0.01029$, $V = 60.28$
(b)   F-4, $n_D = 1.61644$, $n_F - n_C = 0.01684$, $V = 36.61$

In our predesign we shall aim at the following set of thin-lens residuals, hoping that these will give a well-corrected system after suitable thicknesses have been inserted:

$f' = 10$	Petzval sum $= 0.035$
$y_a = 1.111111$	chromatic aberration $= -0.02$
$u_a = 0$	lateral color $= 0$
$u'_0 = 0.111111$	spherical aberration $= -0.08$
$u_{\text{pr. }a} = -0.364(\tan 20°)$	coma$'_s = +0.0025$
$K = S_1/S_2 = 1.0$	ast$'_s = -0.09$

with $y_a = 1.111111$, $y_b = 0.888888$, and $y_c = 0.999999$. Solving the three Schwarzschild equations for the three powers gives

$$\phi_a = 0.192227, \qquad \phi_b = -0.291104, \qquad \phi_c = 0.156285$$

The paraxial ray and the paraxial principal ray passing through the middle of the negative lens have the values shown in the following table:

ϕ $-d$	ϕ_a		ϕ_b		ϕ_c
		$-S_1$		$-S_2$	
		Paraxial ray			
y	y_a		y_b		y_c
u	u_a		u_b	u_c	u'_0
		Paraxial principal ray			
y_{pr}	y_{pr_a}		$y_{pr_b} = 0$		y_{pr_c}
u_{pr}	u_{pr_a}		u_{pr_b}	u_{pr_b}	

Inspection of this table shows that, for the paraxial ray,

$$u_a = 0, \qquad u_b = u_a + y_a\phi_a, \qquad u_c = u_b + y_b\phi_b$$
$$S_1 = (y_a - y_b)/u_b, \qquad S_2 = (y_b - y_c)/u_c$$

Substituting the numerical values of our example gives

$$u_a = 0, \qquad u_b = 0.2135856, \qquad u_c = -0.0451736$$
$$S_1 = 1.040436, \qquad S_2 = 2.459647$$

whence $K = S_1/S_2 = 0.423002$. Now it is found that K varies almost linearly with y_b, and a couple of trials tells us that $\partial K/\partial y_b = -46.0$. Thus retaining the previous $y_a = 1.111111$ and $y_c = 0.999999$, we find that with $y_b = 0.876380$ we have

$$\phi_a = 0.153234, \qquad \phi_b = -0.296588, \qquad \phi_c = 0.200775$$
$$u_b = 0.1702602, \qquad u_c = -0.0896636$$
$$S_1 = 1.378661, \qquad S_2 = 1.378709, \qquad K = 0.999965$$

This is virtually perfect, so we return to the thin-lens ray-trace table and we see that for the paraxial principal ray

$$y_{pr_a} = \frac{S_1 u_{pr_a}}{1 - S_1\phi_a} = -0.636244,$$
$$y_{pr_b} = 0, \qquad y_{pr_c} = -y_{pr_a}/K = +0.636266$$

We can now determine the contribution of each element to the lateral color by the relation

$$T_{ch}C = -yy_{pr}\phi/Vu_0'$$

whence

$$T_{ch}C_a = 0.0161736, \qquad T_{ch}C_b = 0, \qquad T_{ch}C_c = -0.0190729$$

with the total $= -0.002899$. To correct this, we must change y_c and repeat the whole process.

Omitting all the intermediate steps, we come to the final solution:

$$y_a = 1.111111, \qquad y_b = 0.861555, \qquad y_c = 0.962510$$

$$\phi_a = 0.1684127, \qquad \phi_b = -0.3050578, \qquad \phi_c = 0.1940862$$

$$u_b = 0.1871252, \qquad u_c = -0.0756989$$

$$S_1 = 1.333632, \qquad S_2 = 1.333639, \qquad K = 0.999995$$

With $u_{pr_a} = -0.364$, we find

$$y_{pr_a} = -0.6260542, \qquad y_{pr_b} = 0, \qquad y_{pr_c} = 0.6260573$$

whence

$$T_{ch}C_a = 0.0174910, \qquad T_{ch}C_b = 0, \qquad T_{ch}C_c = -0.0174616$$

Hence the thin-lens lateral color is $+0.0000294$, which is acceptable.

B. The Thin-Lens Predesign of the Bendings

The bendings of our three thin-lens elements are defined by c_1, c_3, and c_5, respectively. Since the stop is assumed to be in contact with lens (b), the astigmatism contribution of that element is independent of its bending. Our procedure, therefore, is to adopt some arbitrary bending of lens (a) and ascertain its AC^* by the formulas given on p. 207. We find the AC of lens (b) by $-\frac{1}{2}h_\theta^2\phi_b$ and then solve for the bending of element (c) that will make the total astigmatism contribution equal to the specified value of -0.09. Having done this, we go to lens (b) and bend it to give the desired value of the sagittal coma, namely, 0.0025. This will not affect the astigmatism in any way. Finally, knowing the bendings of elements (b) and (c), we can calculate the spherical contributions of all three elements, and plot a point on a graph connecting spherical aberration with the value of c_1. Repeating this process several times with different values of c_1 will enable us to complete the graph and pick off the final solution for any desired value of the thin-lens primary spherical aberration.

The contributions of the thin lens elements to the three aberrations are given by the formulas on p. 207 involving the G sums for spherical aberration and coma. These contributions are quadratics in terms of the bending parameters c_1, c_3, and c_5 as follows:

Lens a

$$SC^* = -23.227833c_1^2 + 11.968981c_1 - 2.011823$$

$$CC^* = 1.454188c_1^2 - 1.361274c_1 + 0.292417$$

$$AC^* = -7.374247c_1^2 + 10.006298c_1 - 3.442718$$

Lens b

$$SC^* = 15.229687c_3^2 + 4.686647c_3 + 1.436793$$

$$CC^* = 0.667069c_3 + 0.095270$$

$$AC^* = 2.020947$$

Lens c

$$SC^* = -15.073642c_5^2 + 5.519113c_5 - 0.937665$$

$$CC^* = -1.089393c_5^2 - 0.130340c_5 + 0.030780$$

$$AC^* = -6.377286c_5^2 - 3.861014c_5 - 0.528703$$

Collecting these expressions, we find that with a given c_1, we first solve for c_5 by the quadratic expression

$$c_5^2 + 0.6054322c_5 + (1.15633c_1^2 - 1.569053c_1 + 0.2917344) = 0$$

Only one of the two solutions is useful; the other represents a freakish lens bent drastically to the left that would exhibit huge zonal residuals.

Knowing c_1 and c_5, we can solve for c_3 for coma correction by

$$c_3 = -2.179967c_1^2 + 2.040679c_1 + 1.633104c_5^2 + 0.1953917c_5 - 0.6235753$$

Finally, knowing all three parameters c_1, c_3, and c_5, we can calculate the spherical aberration by

$$LA' = -23.227833c_1^2 + 11.968981c_1 + 15.229687c_3^2$$
$$+ 4.686647c_3 - 15.073642c_5^2 + 5.519113c_5 - 1.512695$$

Taking a series of values for c_1 we find

c_1	c_3	c_5	Primary spherical aberration
0.2	−0.308020	−0.042985	−0.311751
0.25	−0.238049	0.043543	−0.013077
0.3	−0.168838	0.105388	0.044583
0.35	−0.100863	0.152719	0.004574
0.4	−0.060233	0.189738	−0.164065

Thus all three lenses are bending to the right together. The spherical sums are plotted on a graph (Fig. 169), from which we can pick off the desired c_1

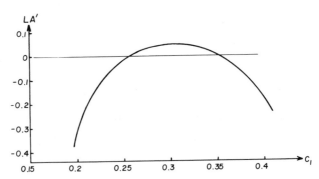

FIG. 169. Relation between c_1 and primary spherical aberration, after correcting field by c_5 and coma by c_3.

values for our residual of −0.08. There are obviously two solutions, at

$$c_1 = 0.2314, \qquad c_1 = 0.3780$$

We shall follow up only the left-hand solution since the right-hand solution has more steeply curved surfaces and is likely to exhibit larger zonal residuals. For the left-hand solution then, we have the following thin-lens curvatures:

$$c_1 = 0.2314, \qquad c_3 = -0.264746, \qquad c_5 = 0.015190$$

$$c_2 = -0.040098, \qquad c_4 = 0.230124, \qquad c_6 = -0.297695$$

C. Calculation of Real Aberrations

After selecting suitable thicknesses from a scale drawing, scaling the lenses up or down to restore their exact thin-lens powers, and calculating the air spaces to maintain the thin-lens separations between adjacent principal points, we obtain the following thick-lens system:

c	d	n_D
0.2326236		
	0.4	1.62031
−0.04031		
	1.051018	
−0.2617092		
	0.25	1.61644
0.227485		
	0.986946	
0.0152285		
	0.45	1.62031
−0.2984403		

with $f' = 10.00$, $l' = 8.649082$, LA' $(f/4.5) = 0.01267$, LZA $(f/6.3) = -0.01051$, OSC $(f/4.5) = -0.001302$, $Ptz = 0.03801$.

Field (deg)	X'_s	X'_t	Distortion (%)	Lateral color
24	−0.0386	−0.4338	1.98	0.00195
20	−0.0639	−0.0798	1.09	0.00055
14	−0.0488	+0.0192	0.42	−0.00021

The lateral color correction is evidently about right. Figure 170 shows

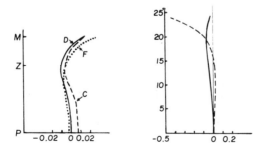

Fig. 170. Aberrations of final triplet lens ($f/4.5$).

the plotted spherochromatism graphs, which show that both the spherical and chromatic aberrations are also about right. The astigmatic fields are also plotted in Fig. 170, where it can be seen that a slight change in the thin-lens Ast_s in the positive direction might be an improvement.

As far as coma is concerned, we have plotted the graphs of H' against Q_1 for sets of oblique rays at three different obliquities in Fig. 171. The points

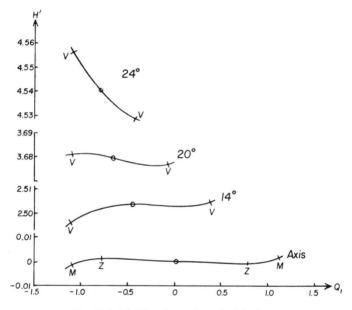

FIG. 171. Meridional ray plots of triplet lens.

VV on each graph represent the limiting vignetted rays, which enter and leave the lens at the initial marginal aperture height of 1.1111, assuming that the front and rear apertures of the lens are limited to that figure (Fig. 172).

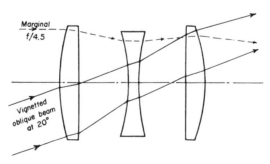

FIG. 172. Final triplet design, showing $f/4.5$ marginal ray and limiting rays of vignetted 20° beam.

When the obliquity is increased the vignetting becomes accentuated, and the graphs become shorter at the upper (right-hand) ends. The principal ray along which the astigmatism was calculated is indicated in each case. The slope of the graph at the principal-ray point is, of course, an indication of the X'_t for that obliquity. To improve the distortion, the design could be repeated with a different value of K, say, 0.9 or 1.1.

On the whole, this seems to be a pretty good design, typical of many triplets using these common types of glass. For a good lens at a higher aperture such as $f/2.8$, for example, it would be highly desirable to use glasses with much higher refractive indices, such as a lanthanum crown and a dense flint. A search through the patent files will reveal many triplet designs for use at various apertures and angular fields.

CHAPTER 14

Mirror and Catadioptric Systems

Curved mirrors, concave or convex, have often been used as image-forming systems, either alone or in combination with lens elements. Historically, most large astronomical telescopes have used a concave mirror to form the primary image, which is then relayed and magnified by either a second concave mirror (Gregorian) or a convex mirror (Cassegrain), although the objectives of small telescopes are generally achromatic lenses. Very small aspheric or spherical mirror systems have occasionally been used as microscope objectives. A single mirror used alone must generally be aspheric to correct the spherical aberration, but by combining two or more mirrors, with perhaps some lens elements also, it is possible to secure good aberration correction using only spherical surfaces. Summaries of reflecting and catadioptric systems have been given by Villa[1] and Gavrilov.[2]

I. COMPARISON OF MIRRORS AND LENSES

Mirrors have many advantages over lenses, principally as follows:

1. A mirror can be made of any size and of any material, even metal, provided it is capable of a high polish. Since good optical glass blanks cannot be made in diameters greater than about 20 in., all optical systems larger than that must be mirror systems. Often a mirror is used in conjunction with lens elements for aberration correction; such systems are called "catadioptric."

2. Mirrors have no chromatic aberrations of any kind; hence a mirror can be focused in the visible and used in any wavelength region in the UV or IR if desired. Also, mirrors exhibit no selective absorption through the spectrum as lenses do, but it must be noted that it is difficult to form mirror coatings that reflect well in the extreme ultraviolet.

3. A mirror has only one-quarter the curvature of a lens having the same power; hence mirrors can have a high relative aperture without the intro-

---

[1] J. Villa, Catadioptric lenses, *Spectra* 1, 57 (Mar.–Apr.), 49 (May–June) (1968).

[2] D. V. Gavrilov, Optical systems using meniscus lens mirrors, *Sov. J. Opt. Technol.* (*Engl. transl.*) 34, 392 (May–June 1967).

duction of excessive aberration residuals. The Petzval sum of a concave mirror is actually negative.

4. By the use of several mirrors in succession, it is often possible to fold up a system into a very compact space.

On the other hand, mirrors have many features that are disadvantageous by comparison with lenses:

1. There will generally be an obstruction in the entering beam, causing a loss of light and a worsening of the diffraction image. This obstruction may be a secondary mirror or an image receiver, and if the angular field is wide the obstruction may block off nearly all of the incident light.

2. Since all the power is in one mirror surface, that surface must conform extremely closely to the desired shape, for even a slight distortion of the surface by the action of gravity or by temperature variations may cause a severe loss of definition. Flexure of a lens causes merely a trivial change in the aberrations, but flexure in a mirror changes the image position and alters the image quality drastically. The problem of mounting a large mirror without any flexural distortion is a very difficult one.

3. The angular field of a mirror system is generally quite small. It can be increased by the addition of one or more lens elements, but then many of the advantages of a mirror are lost.

4. In most reflective systems it is unfortunately possible for light from an object to proceed directly to the image without striking the mirrors. This must be prevented by the use of suitable baffles if the system is to be used in daylight. No baffles are needed in astronomical instruments since the overall sky brightness is very low at night.

II. RAY TRACING A MIRROR SYSTEM

If an optical system contains spherical mirrors, the standard ray-tracing procedure can be readily modified. The surfaces are listed in the order in which they are encountered by the light, with the usual sign convention that radii are regarded as positive if the center of curvature lies to the right of the surface. The separations d, the refractive indices n, and the dispersions Δn are entered as positive quantities if the light is traveling from left to right, but negative if the light is proceeding from right to left. The system should be oriented in such a way that the final imaging rays are moving from left to right so that the image-space index is positive. It may, therefore, be necessary in some cases to regard the object-space index as negative; if this presents difficulties a fictitious plane mirror can be inserted in front of the system to reverse the direction of the incident light.

As an example we will trace a paraxial ray and an $f/1$ marginal ray

TABLE XIII

RAY TRACE THROUGH A CATADIOPTRIC SYSTEM

	init			(mirror)		
c		0.20	0.143	0.1	0.6079	0
d		−0.35	−4.0	5.286	0.6	
n	−1.0	−1.545	−1.0	1.0	1.545	

Paraxial

ϕ	−0.1090000	0.0779350	0.2	0.3313066	0
$-d/n$	−0.2265372	−4.0	−5.286	−0.3883495	
y	2.0	2.049385	2.282510	0.177515	0.000026
nu	−0.2180000	−0.0582812	0.3982208	0.4570327	0.4570327

Marginal (f/1)

Q	2.188	2.298947	2.587155	0.1870214	0.0004700
Q'	2.241196	2.289509	2.463413	0.1945020	0.0004302
l	25.9509	9.4244	10.6206	−18.4788	−18.8679
l'	16.4535	14.6544	−10.6206	−11.8382	−29.9757
$\sin U'$	0.1650029	0.0744115	0.4306454	0.3233869	0.4996328
U'	9.4974	4.2674	25.5085	18.8679	29.9757

Paraxial $l' = 0.000057$　　$f' = 4.376054$

Marginal $L = 0.000861$　　$F' = 4.379217$

$LA' = 0.000804$

through a Gabor system (see Table XIII). This system has a negative corrector lens in front, a concave mirror and a positive field-flattener, the light entering from infinity in a right-to-left direction. It should be noted that in the paraxial trace, the sign of the product nu depends on both the sign of n and on the sign of u.

III. SINGLE-MIRROR SYSTEMS

A. A SPHERICAL MIRROR

A spherical mirror with an object point at its center of curvature is a perfect optical system having no aberrations of any kind. If the object point is displaced from the center of curvature, the paraxial image point moves in the opposite direction along a straight line joining the object point to the center of curvature (Fig. 173). Because the aperture stop is at the mirror, the

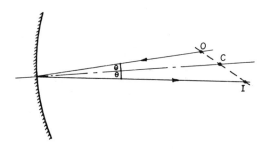

FIG. 173. The line joining object and image passes through the center of curvature of a spherical mirror.

system is symmetrical, and for small object displacements there will be no coma. However, some astigmatism will be introduced, the sagittal image coinciding with the paraxial image at the Lagrangian image point while the tangential image is somewhat backward-curving.

It should be remarked that the focal length of a single spherical mirror is exactly half the radius of curvature; the principal points coincide at the vertex of the mirror, while the nodal points coincide at the center of curvature. Because the refractive indices of the overlapping object and image spaces are equal and opposite, the two focal lengths have the same sign, and the distance from the principal point to the nodal point is equal to twice the focal length. This applies to all reflective and catadioptric systems having an odd number of mirrors. With an even number of mirrors, the outside refractive indices have the same sign, and the ordinary rules for a lens system apply.

If a spherical mirror is used with a distant object, undercorrected spherical aberration and overcorrected OSC appear at the focus. The magnitude of these aberrations is seen from the ray diagram in Fig. 174. Here

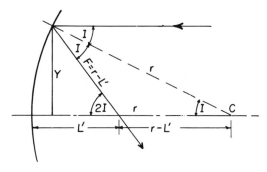

FIG. 174. The spherical aberration of a spherical mirror.

$$\sin 2I = Y/(r - L'), \qquad \sin I = Y/r$$

Hence

$$\frac{Y}{r - L'} = 2\left(\frac{Y}{r}\right)\left[1 - \left(\frac{Y}{r}\right)^2\right]^{1/2}$$

From which we find

$$L' = r - r^2/2(r^2 - Y^2)^{1/2}, \qquad F' = r - L'$$

A few points calculated for a mirror with radius 20 and focal length $f' = 10$ are given in the following tabulation:

Y	L'	$LA' = L' - l'$	$F' = r - L'$	OSC	Aperture
0.1	9.999875	−0.000125	10.000125	0.000025	$f/50$
0.2	9.999500	−0.000500	10.000500	0.000100	$f/25$
0.5	9.996874	−0.003126	10.003126	0.000625	$f/10$
1.0	9.987477	−0.012523	10.012523	0.002508	$f/5$
2.0	9.949622	−0.050378	10.050378	0.010127	$f/2.5$

In this case the standard OSC formula becomes simplified to $(F'/L' - 1)$ because $l'_{pr} = 0$ and $l' = f'$. It should be noted that the spherical aberration is purely primary for apertures less than about $f/6$, and that by $f/5$ the OSC has already reached Conrady's tolerance of 0.0025 for telescope objectives. For apertures under $f/10$ a single spherical mirror is often as good as a parabolic mirror and it is, of course, very much cheaper to manufacture.

B. A PARABOLIC MIRROR

To determine the correct form for a concave mirror to be free from spherical aberration, we consider a plane wave front reaching the mirror from an axial object point at infinity (Fig. 175). In this diagram the entering

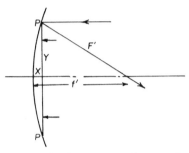

FIG. 175. Reflection of a plane wave PP by a parabolic mirror.

plane wave is PP, and while the axial portion of the wave is traveling a distance $X + f'$, the marginal part of the wave travels a distance F'. Hence

$$F' = X + f' = [Y^2 + (f' - X)^2]^{1/2}$$

whence

$$Y^2 = 4f'X$$

This is clearly the equation of a parabola with vertex radius equal to $2f'$.

This property of a parabolic mirror has been known for centuries, and it is the form given to the primary mirror in most reflecting telescopes. However, this mirror suffers from high OSC. The focal length F' of a marginal ray is equal to $[Y^2 + (f' - X)^2]^{1/2}$ and it increases as Y increases in the same manner as in a spherical mirror with a distant object. The coma corresponding to an object subtending an angle U_{pr} is given by

$$(F' - f') \tan U_{pr} = \{[4f'X + (f' - X)^2]^{1/2} - f'\} \tan U_{pr} = X \tan U_{pr}$$

If the aperture of the mirror is small, we can write $X = Y^2/2r$, and the sagittal coma becomes simply

$$\text{coma}_s = h'/16(f\,\text{number})^2 \qquad \text{or} \qquad OSC = 1/16(f\,\text{number})^2$$

The same result can be derived from the primary coma expression on p. 207. Thus at the prime focus of the Palomar telescope, for example, where the f number is 3.3, the sagittal coma at a point only 20 mm off-axis has reached a magnitude of 0.115 mm. It will be found that this OSC is the same whether the mirror is a sphere or a parabola, but of course the spherical aberration is quite different in the two cases.

If the obstruction caused by the image receiver is undesirable, a so-called off-axis parabola may be used (Fig. 176). The only practical way to con-

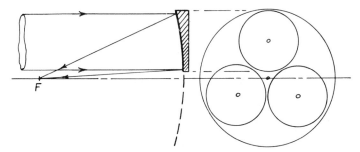

FIG. 176. Cutting three off-axis parabolic mirrors from one large paraboloid.

struct such a mirror is to make a large on-axis mirror and cut as many off-axis mirrors from it as are needed. Such mirrors are used in mirror monochromators of the Wadsworth type, and as schlieren mirrors for wind tunnel applications.

C. An Elliptical Mirror

As mentioned previously (p. 36), the equation of a conic section is

$$X = cY^2/\{1 + [1 - c^2Y^2(1 - e^2)]^{1/2}\}$$

where c is the vertex curvature and e the eccentricity. For an ellipse, e lies between 0 for a circle and 1 for a parabola. If a and b are the major and minor semiaxes of the ellipse, respectively, then

$$e = [(a^2 - b^2)/a^2]^{1/2}, \qquad a = 1/c(1 - e^2), \qquad b = 1/c(1 - e^2)^{1/2} = (a/c)^{1/2}$$

In terms of the two semiaxes, the vertex curvature is $c = a/b^2$.

A concave elliptical mirror has the interesting optical property of two "foci," which are such that an object point located at one is imaged at the other without aberration. The two "focal lengths," i.e., the distances from the mirror vertex to the two foci, are

$$f_1 = a(1 - e), \qquad f_2 = a(1 + e)$$

Hence

$$e = (f_2 - f_1)/(f_2 + f_1), \qquad a = \tfrac{1}{2}(f_1 + f_2), \qquad b = (f_1 f_2)^{1/2}$$

All optical paths from one focus to the other via a point on the ellipse are equal, but the magnification along each path is given by the ratio of the two sections of the path, and hence it varies greatly from point to point along the curve. This leads to heavy coma for an off-axis object point.

If the ellipse is turned so that the vertex is at the middle of the long side, we have an oblate spheroid, and then the "conic constant" $1 - e^2$ is greater than 1.0. This situation seldom arises, however, since an oblate spheroid is stronger than a sphere at the margin, and so it has worse spherical aberration.

To draw an ellipse, we first construct the two auxiliary circles on the major and minor axes as shown in Fig. 177a, and we draw any transversal

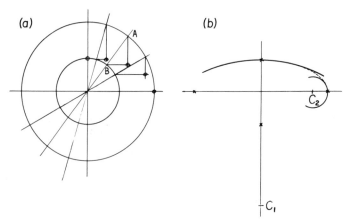

FIG. 177. How to draw an ellipse.

through the midpoint. If this crosses the two circles at A and B, respectively, then the point of intersection of a vertical line through A and a horizontal line through B is a point on the ellipse. By running several such transversals, enough points can be plotted to enable the ellipse to be filled in by use of a French curve. A simpler but less accurate procedure is to calculate the two vertex radii b^2/a and a^2/b and draw arcs with these radii through the ends of the semiaxes as in Fig. 177b. These arcs almost meet, and the small gaps can be readily filled in with a French curve. A combination of both methods is probably the best procedure.

D. A HYPERBOLIC MIRROR

The eccentricity of a hyperbola is greater than 1.0, so that the conic constant $1 - e^2$ is negative. A hyperbola has two branches, and a hyperbolic mirror is formed usually by rotating the hyperbola about its longitudinal axis, only one branch being utilized. This may be either a convex or a concave mirror. If convex, then any ray directed toward the inside "focus" will be reflected through the outside "focus," the two focal lengths being

$$f_1 = a(1 - e), \qquad f_2 = a(1 + e)$$

where a is the distance along the axis from the mirror vertex to the midpoint of the complete hyperbola (Fig. 178). The separation of the vertices of the

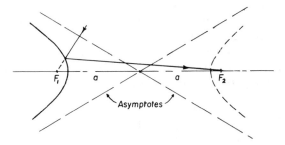

FIG. 178. A convex hyperbolic mirror. ($a = 38$, $r = -12.8$, $e = 1.156$, $f_1 = -5.93$, $f_2 = 81.93$).

two hyperbolic branches is, of course, $2a$. The vertex radius is $a(1 - e^2)$, so that f_1 and f_2 satisfy the ordinary mirror conjugate relation

$$1/f_1 + 1/f_2 = 2/r$$

A convex hyperbolic mirror is used in the Cassegrain telescope, and a concave hyperbola is used in the Ritchey–Chrétien arrangement.

IV. SINGLE-MIRROR CATADIOPTRIC SYSTEMS

It was suggested by F. E. Ross in 1935[3] that it might be possible to remove the coma from a parabolic mirror by inserting an air-spaced doublet lens of approximately zero power into the imaging light beam, at a position fairly close to the image to keep the lens small. Since the lens was to be a thin achromat of zero power, the same glass could be used for both elements. Ross found that it is impossible to correct all three aberrations, spherical, coma, and field curvature, simultaneously, and so he worked with coma and field, letting the spherical aberration fall where it would. However, by greatly increasing the lens powers, it is possible to design an aplanat corrected for spherical and coma, but the field is then decidedly inward-curving. Examples of both systems will be given here.

A. A FLAT-FIELD ROSS CORRECTOR

Assuming a parabolic mirror of vertex radius 200 and focal length 100, the spherical aberration is, of course, zero, and the marginal focal length at

---

[3] F. E. Ross, Lens systems for correcting coma of mirrors, *Astrophys. J.* **81**, 156 (1935).

$f/3.33$ is found to be 100.5625. The *OSC* of the mirror is therefore 0.005625 when the stop is at the mirror so that $l'_{pr} = 0$. Tracing a principal ray entering the mirror vertex at 0.5°, we find that $X'_s = 0$ and $X'_t = -0.00762$. The Petzval sum is -0.01, giving $X'_{Ptz} = +0.00381$. The tangential astigmatism is exactly three times the sagittal astigmatism at this small obliquity.

We will follow through the design of a Ross corrector to be inserted at a distance of 90 from this parabolic mirror. To avoid vignetting at a field of 0.5° the diameter of the corrector must be about 5.0. The entering data for the three rays are

Marginal: $U = 8.57831°$, $Q = 1.49161$
Paraxial: $u = 0.15$, $y = 1.50$
Principal: $U_{pr} = -0.5°$, $Q_{pr} = 0.7853882$

We will start with the following setup. The glass is K-3 with $n_e = 1.52031$ and $V_e = 59.2$:

	c	d	n
	0		
		0.3	1.52031
	0.1		
		0.089228	
	0.07		
		0.65	1.52031
$(D-d)$	-0.036683		

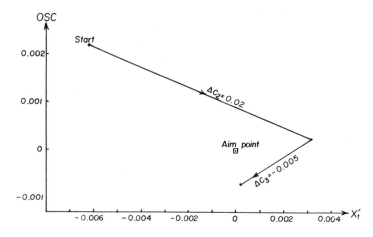

FIG. 179. Double graph for a flat-field Ross corrector.

with $f' = 97.5837$, $l' = 9.17044$, $LA' = -0.06648$, $OSC = 0.00221$, Ptz $= -0.00771$; for $0.5°$: $X'_s = -0.00024$, $X'_t = -0.00636$, distortion $= +0.09\%$. The plane front face is arbitrary and will be retained throughout. The central air space has been calculated to permit the two lenses to be in edge contact at a diameter of 4.8, and the last radius is calculated by the $D - d$ method for perfect achromatism. The dispersion of the glass need not be known since both elements are made of the same material.

The two variables that will be used to achieve coma correction and a flat tangential field are, of course, c_2 and c_3. Making small changes in these variables permits us to plot the double graph shown in Fig. 179. After a few trials, the final system was as follows:

c	d	n
0		
	0.3	1.52031
0.1169		
	0.149348	
0.0670		
	0.65	1.52031
$(D - d)$ -0.0576113		

with $f' = 97.4760$, $l' = 9.18666$, $LA' = -0.11509$, $OSC = -0.00001$, Ptz $= -0.00736$; for $0.5°$: $X'_s = +0.00153$, $X'_t = -0.00056$, distortion $= +0.19\%$. The passage of axial and oblique rays through this system is shown in Fig. 180.

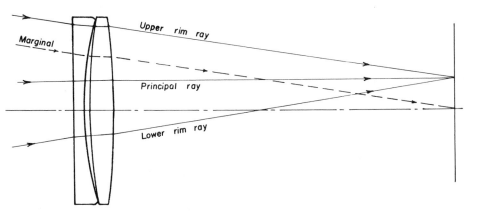

FIG. 180. Path of rays through Ross corrector.

The slightly backward-curving sagittal field shown in this table could probably be corrected by the use of a somewhat higher refractive index for the negative element, but this possibility was not explored. The major problem is, of course, the large residual of spherical aberration, which could be removed only by the use of an aspheric surface. Some recent workers have managed to correct all three aberration by means of three or more elements.

B. An Aplanatic Parabola Corrector

By making both elements considerably stronger, it is possible to correct the spherical aberration and *OSC*, and thus design an aplanatic corrector, but only at the expense of a considerable inward field curvature. The thickness of the positive element must be increased, and the central air space must be held at some fixed value because the adjacent surfaces are almost identical.

For a starter we may consider the following setup:

c	d	n
-0.1		
	0.3	1.52031
0.1		
	0.1	
0.1		
	1.1	1.52031
$(D - d)$ -0.1095215		

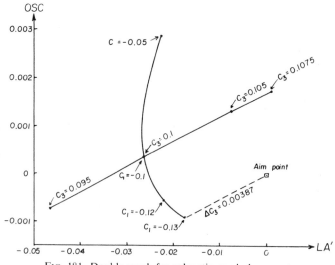

FIG. 181. Double graph for aplanatic parabola corrector.

with $f' = 97.5847$, $l' = 9.33076$, $LA' = -0.02649$, $OSC = 0.00037$, Ptz $= -0.00674$; for $0.5°$: $X'_s = -0.0139$, $X'_t = -0.0469$, distortion $= -0.12\%$. We will now hold the second surface curvature arbitrarily at 0.1, and vary the other curvatures c_1 and c_3 to plot a double graph (Fig. 181). The graph for changes in c_1 is found to be decidedly curved, which is not surprising since changes in c_1 represent both a bending and a power change, whereas changes in c_3 are a pure bending. A few trials give us the following final setup:

c	d	n
-0.13		
	0.3	1.52031
0.1		
	0.1	
0.10387		
	1.1	1.52031
$(D-d)$ -0.1322352		

with $f' = 98.7691$, $l' = 9.58664$, $LA' = 0.00001$, $OSC = 0.00002$, Ptz $= -0.00791$; for $0.5°$: $X'_s = -0.0196$, $X'_t = -0.0654$, distortion $= -0.21\%$. This system would be extremely heavy if made in a large size, and the inward tangential field curvature would be obviously undesirable, being about nine times as great as for the mirror alone. Other values of c_2 could, of course, be tried, but the result is likely to be similar to this in performance.

C. THE MANGIN MIRROR

The French engineer Mangin[4] in 1876 proposed replacing the parabolic mirror in a searchlight by a more easily manufactured spherical mirror, with a thin meniscus-shaped negative lens in contact with the mirror to correct the spherical aberration (Fig. 182). The design procedure is simple since there is only one degree of freedom, namely, the outside radius of the lens, because the mirror radius determines the focal length of the system. Using K-4 glass $(n_e = 1.52111$, $V = 57.64)$, a few trials give the following setup:

c	d	n
0.0981		
	0.3	1.52111
(mirror) 0.06544		

[4] A. Mangin, *Mémorial de l'officier du génie (Paris)* **25** (2), **10**, 211 (1876).

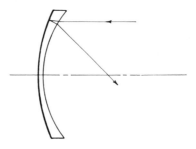

FIG. 182. A typical Mangin mirror.

with $f' = 10.0155$, $l' = 9.82028$, LA' $(f/3) = 0.00001$, LZA $(f/4.2) = -0.00008$, OSC $(f/3) = 0.00307$. The OSC is less than half that of a parabolic mirror of the same focal length and aperture, but the chromatic aberration from F to C is found to be 0.0564, while the zonal spherical aberration is negligible. The next step, therefore, is to achromatize the system.

If we replace the simple negative lens by an achromat using the following glasses:

1. F-4: $n_C = 1.61164$, $n_e = 1.62058$, $n_F = 1.62848$
2. K-4: $n_C = 1.51620$, $n_e = 1.52111$, $n_F = 1.52524$

with the flint element adjacent to the mirror, we may start with a plano interface:

	c	d	n
	0.1		
		0.2	1.52111
	0		
		0.3	1.62058
(mirror)	0.062		

with $f' = 9.8332$, $l' = 9.52178$, LA' $(f/3) = -0.02094$, zonal chromatic aberration $F - C = -0.03326$.

To plot a double graph for the simultaneous correction of spherical and zonal chromatic aberrations, we make trial changes of $\Delta c_1 = 0.01$ and $\Delta c_2 = 0.01$, respectively. The graph so obtained indicates that we should try $c_1 = 0.1021$ and $c_2 = 0.01625$. This is a great improvement, since $LA' = -0.00705$ and the zonal chromatic aberration $L'_{ch} = -0.00560$. A few further small adjustments gave the following final system:

	c	d	n
	0.10636		
		0.2	1.52111
	0.01489		
		0.3	1.62058
(mirror)	0.062		

with $f' = 10.8324$, $l' = 10.52127$, $LA' = -0.00007$, $OSC = 0.00215$, zonal chromatic $= 0.00002$. By tracing other rays the spherochromatism curves can be plotted as in Fig. 183. It can be seen that the aberration residuals are

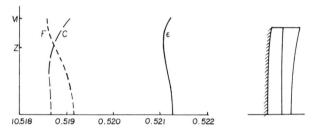

FIG. 183. Spherochromatism of an achromatic Mangin mirror.

very small, the chief residual being the ordinary secondary spectrum typical of a negative achromat. This system is practical if made in small sizes, but for large systems a parabolic mirror would be preferable.

D. THE BOUWERS–MAKSUTOV SYSTEM

During World War II, Bouwers[5] and Maksutov[6] independently proposed the use of a monocentric catadioptric system to cover a wide angular field. This system consisted of a spherical mirror and a thick corrector plate, all three surfaces having a common center C located at the middle of the stop. Such a system has no coma or astigmatism and the image lies on a spherical surface, also concentric about C. The corrector lens can be located either in front of or behind the stop, and it may be thin and strongly curved, or thick and less strongly curved (Fig. 184). For any given front radius the thickness can be adjusted to eliminate the marginal spherical aberration, but the zonal residual will vary with the thickness.

---

[5] A. Bouwers, "Achievements in Optics." Elsevier, New York, 1946.
[6] D. D. Maksutov, New catadioptric meniscus systems, *J. Opt. Soc. Am.* **34**, 270 (1944).

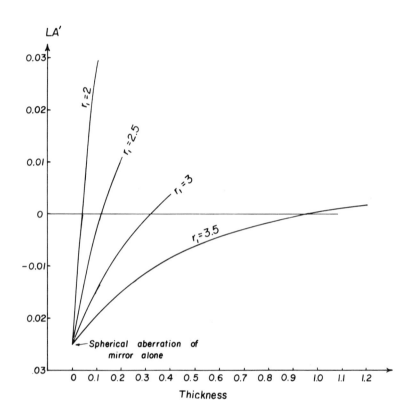

FIG. 184. Graphs connecting the marginal spherical aberration with the thickness of the corrector plate for various values of r_1 (mirror radius = 10).

Although the angular field of this monocentric system is theoretically unlimited, the obstruction caused by the receiving surface increases as the field is widened to the point where eventually no light at all will enter the system. To reduce this effect the relative aperture must be increased as the field is widened, unless, of course, the receiver is a narrow strip crossing the middle of the aperture.

Figure 185a shows the zonal spherical aberration of four examples of Maksutov correctors used with a mirror of radius 10.0, the marginal aberration at $f/2.5$ being corrected in each case by using a suitable thickness for the corrector. The four cases are as follows:

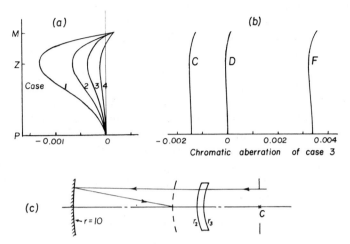

FIG. 185. Bouwers–Maksutov systems. (a) Spherical aberration for four values of r_1. (b) Chromatic aberration of case 3. (c) Ray diagram of case 3.

Case	r_1	Thickness	r_2	Focal length	Back focus
1	2.0	0.040	2.04	4.9172	5.0828
2	2.5	0.121	2.621	4.8463	5.1537
3	3.0	0.320	3.32	4.7386	5.2614
4	3.5	0.950	4.45	4.5260	5.4740

For the third case, the chromatic aberration is shown graphically in Fig. 185b and a scale drawing of the system in Fig. 185c.

The chromatic aberration of the Bouwers–Maksutov system is decidedly large and could be serious. It can be removed by achromatizing the corrector lens, but then the system is no longer monocentric, and the angular field immediately becomes limited. However, if only a narrow field is desired, then achromatizing the corrector is quite a satisfactory procedure.

E. The Gabor Lens

In 1941 Dennis Gabor,[7] the inventor of the hologram, patented a catadioptric system that resembled the Bouwers–Maksutov except that it was not monocentric; it was much more compact and covered a narrow field at a

[7] D. Gabor, Brit. Patent 544,694, filed Jan. 1941.

high relative aperture. Actually, the example shown by Gabor was not achromatic, but the example to be given here is.

In the absence of a field flattener, if the negative front collector lens has a zero $D - d$ sum, the system will obviously be achromatic. Thus the only requirement for achromatism is that the length D measured along the marginal ray inside the corrector lens should be equal to the axial thickness of that lens. To secure this condition it is helpful to make the lens as thick as practical, and to use a glass of moderately high refractive index such as a barium crown. The front radius is then chosen for spherical aberration correction when used with a spherical mirror, and the second radius is found by the ordinary $D - d$ method. Placing the stop at the front surface, the field is backward-curving, and the Petzval sum is negative. The following system was the result of a few easy trials:

c	d	n_c
0.25		
	0.4	1.61282
0.2347439		
	8.0	(air)
(mirror) 0.06		

with $f' = 8.0383$, $l' = 8.59345$, LA' $(f/1.6) = 0.00925$, LZA $(f/2.3) = -0.00420$, OSC $(f/1.6) = 0.00327$, Petzval sum $= -0.1258$. The fields at an obliquity of $1°$ were

$$X'_s = 0.00104, \qquad X'_t = 0.00064, \qquad \text{distortion} = -0.012\%$$

As Gabor indicated in his patent, the negative Petzval sum can be easily eliminated by the addition of a positive field flattener close to the image plane. This lens may conveniently be plano-convex, although it may require a slight bending to flatten the tangential field. A possible starting system with such a field lens is as follows:

	c	d	n_c	Glass
(as before)	$\begin{cases}0.25\\0.2347439\end{cases}$	0.4	1.61282	SK-1
		8.0		
(mirror)	0.06			
		8.0		
(field flattener)	$\begin{cases}0.37162\\0\end{cases}$	0.1	1.51173	K-1

with $f' = 7.2231$, $l' = 0.46712$, LA' $(f/1.6) = -0.00651$, OSC $(f/1.6) = -0.00348$; for $1°$: $X'_s = 0.00008$, $X'_t = 0.00024$, distortion = 0.025%. This field flattener has introduced a small amount of negative $D - d$ sum, which is easily removed by a small change in the radius of the second surface of the correcting lens. The front surface was also strengthened slightly to remove the small residual of spherical undercorrection caused by the field flattener. The final system is as follows:

	c	d	n_e
	0.251		
		0.4	1.61282
	0.2348373		
		8.0	
(mirror)	0.06		
		8.0	
	0.37264		
		0.1	1.51173
	0		

with $f' = 7.1775$, $l' = 0.49554$, LA' $(f/1.6) = 0.00501$, LZA $(f/2.3) = -0.00219$, OSC $(f/1.6) = -0.00468$, $Ptz = 0$; for $1°$: $X'_s = 0.00009$, $X'_t = 0.00028$, distortion = 0.025%.

To investigate the coma, it is necessary to make a meridional ray plot for the $1°$ beam. This is shown in Fig. 186 next to the corresponding plot for the

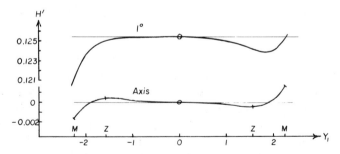

FIG. 186. Meridional ray plot for Gabor lens with space 8.0.

axial beam. It is clear that there is an excess of negative coma present, which can be removed by shifting the corrector lens along the axis. As this has very little effect on the aberrations, it is advisable to make a large shift, say from 8.0 to 6.0. This causes a slight overcorrection of the spherical aberration, requiring a weakening of the front surface and a recalculation of the $D - d$

sum. These changes lead to the following results:

	c	d	n_e
	0.246		
		0.4	1.61282
	0.2303761		
		6.0	
(mirror)	0.06		
		8.0	
	0.37264		
		0.1	1.51173
	0		

with $f' = 7.2492$, $l' = 0.49128$, LA' $(f/1.6) = 0.00314$, LZA $(f/2.3) = -0.00299$, OSC $(f/1.6) = -0.00163$. $Ptz = 0.000204$; for $1°$: $X'_s = 0.00003$, $X'_t = 0.00009$, distortion $= 0.026\%$. The $(D - d)\,\Delta n$ in the two lenses is ± 0.0000343. To complete the study, the $1°$ meridional ray plot was drawn

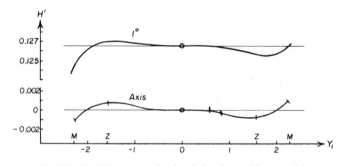

FIG. 187. Meridional ray plot for Gabor lens with space 6.0.

(Fig. 187). The improvement over the previous setup is obvious.

Although this system is well corrected, mechanically something must be done to keep the imaging rays clear of the corrector lens. A possible arrangement is shown in Fig. 188, using a hole in the middle of the corrector lens, but this hole must be quite large for so high an aperture as $f/1.6$. A plane mirror could be employed to reflect the beam out sideways, or back through the middle of the concave mirror.

This Gabor system is unusual in that each of the six degrees of freedom (five radii and one air space) is almost specific for one particular aberration. The front surface controls the spherical aberration and the second the chromatic aberration; the power of the field lens determines the Petzval sum, while its bending controls the field curvature; and finally the central air

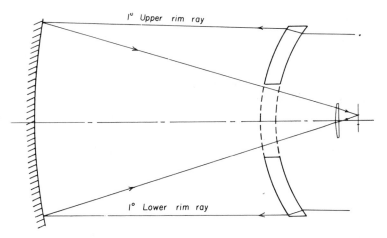

FIG. 188. Final design of $f/1.6$ Gabor system covering $\pm 1°$.

space is used to vary the coma. The mirror radius, of course, determines the focal length. The two remaining aberrations, lateral color and distortion, are usually negligible at such a narrow field, but if the lateral color should be significant, it may be necessary to achromatize the field lens, which would require a further adjustment of the chromatic correction by c_2. The aperture of the Gabor system can be high, but the angular field is small.

F. THE SCHMIDT CAMERA

The Schmidt camera[8] consists of a concave spherical mirror with a thin aspheric corrector plate located at the center of curvature of the mirror. By placing the stop at the corrector plate we automatically eliminate coma and astigmatism, although at high obliquities some higher-order aberrations appear, but the useful field of several degrees is much larger than that of most catadioptric systems. The remaining aberration is spherical, which is corrected by a suitable aspheric surface on the corrector plate. The chromatic aberration is ignored.

The simplest way to derive an expression for the shape of the aspheric surface is to select a neutral zone to represent the minimum point on the aspheric surface, where the plate is momentarily parallel, and let the ray through this neutral zone define the focal point of the system. Tracing a paraxial ray backwards from this focus and performing an angle solve enables us to determine the vertex radius of the aspheric surface. To deter-

---

[8] B. Schmidt, *Mitt. Hamburg Sternw. Bergedorf* 7, 15 (1932).

mine the thickness of the plate at the neutral zone we must equalize the optical paths along the paraxial ray and the neutral-zone ray. We now have three relationships by which three terms of the aspheric polynomial can be found, namely, the vertex curvature, the sag of the neutral zone, and the slope of the surface at the neutral zone, which is zero. If we need greater precision or if we desire more than three terms in the polynomial, we can trace several other rays backwards from the focus and make a least-squares solution for as many terms as we need.

The path of the neutral-zone ray is shown in Fig. 189a. The point C is the

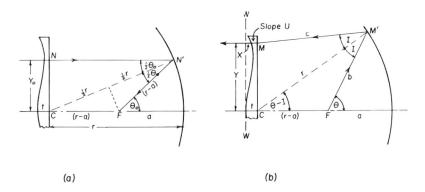

(a)   (b)

FIG. 189. Design of a Schmidt camera. (a) Neutral zone; (b) any other zone.

center of curvature of the concave mirror of radius r. The point F is the focus defined by the intersection of the neutral-zone ray with the axis. The focal length of the system is FN', and the back focal distance is a. If θ_0 is the slope of the neutral-zone ray at the image and Y_0 the incidence height of this ray, then

$$\sin \tfrac{1}{2}\theta_0 = Y_0/r, \qquad a = r - r/2 \cos \tfrac{1}{2}\theta_0$$

The path of any other ray traced backwards from the focus at a starting angle θ is shown in Fig. 189b. The angle of incidence I of this ray at the mirror is given by

$$\sin I = (r - a) \sin \theta/r$$

Assuming for simplicity that the plane side of the corrector plate is at the exact center of curvature of the mirror, the ray path in the air can be calculated by

$$b = FM' = r \sin(\theta - I)/\sin \theta$$

$$c = M'M = r \cos(\theta - I)/\cos(\theta - 2I)$$

The slope of the ray inside the corrector plate is found by

$$\sin U = (1/n) \sin(\theta - 2I)$$

The line WW in Fig. 189b represents a plane wave in the object space, and the optical paths from this wave front to the focus F must be equal along all rays. Along the axis this optical path is evidently $(nt + r + a)$, and along a general ray it is $[b + c + X + n(t - X)/\cos U]$. Equating these paths gives the x coordinate of a point on the asphere as

$$X = \frac{a + r - b - c + nt(1 - \sec U)}{1 - n \sec U}$$

To determine the corresponding height of incidence Y of this ray, we have

$$Y = MC - (t - X) \tan U, \qquad \text{where} \quad MC = c \sin I/\cos(\theta - I)$$

We can apply these formulas to the neutral zone, for which we find

$$b = r - a, \qquad c = r \cos \tfrac{1}{2}\theta_0, \qquad U = 0$$

Example. For an $f/1$ Schmidt with $r = 4.0$, the focal length is about 2 and the marginal ray enters at a height $Y = 1.0$. We may set the neutral zone at an incidence height of 0.85, whence $\sin \tfrac{1}{2}\theta_0 = 0.2125$ and $\theta_0 = 24.5378°$. The focal length of the neutral zone is $F' = 0.85/\sin \theta_0 = 2.046745$ and the back focus is $a = 1.953255$. For the neutral-zone ray we have $b = 2.046745$ and $c = 3.908644$, whence $X_0 = 0.004080$.

We next set the axial thickness of the plate at 0.01, and tracing a paraxial ray backwards from the focus, we solve the vertex curvature of the plate to make the paraxial ray emerge parallel to the axis. In this way we find that the vertex radius should be $R = 45.7416$. The paraxial focal length is 2.046899.

Assuming a three-term polynomial of the form

$$X = AY^2 + BY^4 + CY^6$$

we see that $A = 1/2R = 0.0109310$. For the height of the neutral zone we have

$$X_0 = A(0.85)^2 + B(0.85)^4 + C(0.85)^6 = 0.004080$$

and for the slope of the surface at the neutral zone we have

$$(dX/dY)_0 = 2AY + 4BY^3 + 6CY^5$$
$$= 2A(0.85) + 4B(0.85)^3 + 6C(0.85)^5 = 0$$

Solving these three equations simultaneously gives the three coefficients as

$$A = 0.010931, \qquad B = -0.00681084, \qquad C = -0.00069561$$

Calculating X for several values of Y gives the data needed to plot the shape of the asphere as follows:

Y	X	Y	X	Y	X
0.1	0.000109	0.4	0.001572	0.7	0.003639
0.2	0.000426	0.5	0.002296	0.8	0.004077
0.3	0.000928	0.6	0.003020	0.9	0.004016
				1.0	0.003425

If this curve is plotted, it will be seen that the central bulge is much larger than the curl-up at the rim, so that it might have been better to set the neutral zone a little lower, say at 0.80 instead of 0.85 of the marginal height.

The trivial difference between the zonal and paraxial focal lengths represents an *OSC* of only -0.000075, which is obviously negligible. It could be removed completely by a slight shift of the corrector plate along the axis.

G. SELF-CORRECTED UNIT-MAGNIFICATION SYSTEMS

Two very interesting systems have been proposed for 1:1 imagery, which are automatically corrected for all the primary aberrations.

1. *The Dyson Catadioptric System*[9]

This is a monocentric system (Fig. 190), the object and image lying in the

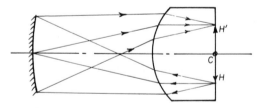

FIG. 190. The Dyson autocollimating system.

same plane on opposite sides of the center of curvature C. A marginal ray from C returns along its own path, thus automatically removing spherical and chromatic aberrations. The radius of curvature of the lens is set at $(n-1)/n$ times the radius of curvature of the mirror, to give a zero Petzval sum. The aperture stop is at the mirror, making a symmetrical system that is automatically corrected for the three transverse aberrations. The seventh

---

[9] J. Dyson, Unit magnification optical system without Seidel aberrations, *J. Opt. Soc. Am.* **49**, 713 (1959).

aberration, astigmatism, is zero near the middle of the field and the sagittal field is flat, but the tangential field bends somewhat backwards at increasing distances out from the axis. A typical system is the following:

c	d	n
0		
	3.434012	1.523
0.2912046		
	6.565988	(air)
(mirror) 0.1		

with

$$l = l' = 0, \qquad m = -1$$

$$H' = 1: \qquad X'_s = 0, \quad X'_t = 0.01460$$

$$H' = 1.5: \qquad X'_s = 0, \quad X'_t = 0.08776$$

As can be seen, the system would be telecentric except for the spherical aberration of the principal ray at the lens surface: the principal ray for $H' = 1.5$ enters at a slope angle of almost 4° in air (2.58° in the glass).

2. The Offner Catoptric System[10]

This monocentric system is similar to the Dyson arrangement, except that a small convex mirror is placed midway between the concave mirror and the object to give a zero Petzval sum, and the beam is reflected twice at the concave mirror (Fig. 191). The aperture stop is at the small convex mirror and the system is virtually telecentric.

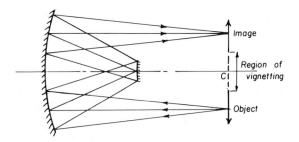

FIG. 191. The Offner autocollimating system.

[10] A. Offner, New concepts in projection mask aligners, Opt. Eng. **14**, 131 (1975).

Because the two mirrors are concentric about C, an object point placed there would be imaged on itself without aberration. However, this is academic because the entire axial beam is blocked out by the secondary mirror. For object points lying off-axis, the vignetting becomes progressively less and finally disappears for object points with H and H' equal to or greater than the diameter of the convex mirror. The symmetry about the stop ensures that coma and distortion are absent. There are, of course, no chromatic aberrations of any kind.

The remaining aberration, astigmatism, is zero for object points near the axis, and the sagittal field is flat, as for the Dyson case. However, the tangential field bends slightly backwards for extraaxial object points.

As an example of this system we may consider the following:

	c	d
Concave	0.1	
		5
Convex	0.2	
		5
Concave	0.1	

with

$$l = l' = 10.0, \qquad m = -1$$
$$H' = 1: \qquad X'_s = 0, \quad X'_t = 0.00205$$
$$H' = 2: \qquad X'_s = 0, \quad X'_t = 0.03519$$

It will be seen that the astigmatism is much smaller than in the Dyson system, and moreover, the long air space between the mirrors and the object plane permits the insertion of plane mirrors to deflect the beam if desired.

V. TWO-MIRROR SYSTEMS

The classical two-mirror systems used in telescopes date from the seventeenth century. They were either of the Gregorian form with a concave parabolic primary mirror and a concave elliptical secondary, or of the Cassegrain form with the same parabolic primary but a convex hyperbolic secondary. The Gregorian form was popular for a hundred years as a small erecting telescope for terrestrial observation. Because of the near impossibility of making an accurate convex hyperboloid, the Cassegrain form only gradually came into use as grinding and polishing techniques were improved. Today Cassegrain telescopes are found in every astronomical observatory.

A. Two-Mirror Systems with Aspheric Surfaces

Suppose we lay out a simple Cassegrain system as shown in Fig. 192. The

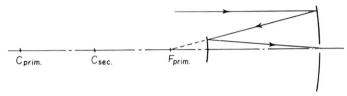

FIG. 192. A simple Cassegrain system.

primary mirror has a radius of curvature equal to 8.0, a focal length of 4.0, and a clear aperture of 2.0 ($f/2$). The secondary mirror has a radius of 3.0 with conjugate distances of -1 and $+3$, forming a final image at the middle of the primary mirror at a magnification of three times. Thus the overall system has a focal length of 12.0 and a relative aperture of $f/6$.

Starting with two spherical mirrors, we find a large residual of undercorrected spherical aberration. This was eliminated in the classical Cassegrain system by making the primary mirror parabolic, of eccentricity equal to 1.0, and the secondary mirror hyperbolic, with an eccentricity of 2.0 in our case. This served to remove the spherical aberration perfectly, leaving an OSC residual of 0.001736. The paths of a marginal ray through each of these systems are shown in Table XIV. For the OSC calculation the stop was assumed to be at the primary mirror, its image being at a distance of $l'_{pr} = -1$ from the secondary mirror.

In the late 1920s Ritchey and Chrétien recognized that the cause of the coma in the classical Cassegrain is that the final U' of the marginal ray is too small, making the marginal focal length F' too long. They therefore suggested departing from the conventional forms of the two mirrors and using shapes that are somewhat flattened at the edge. A few trials show that in our example the eccentricity of the primary mirror should be raised from 1.0 to 1.0368 (a weak hyperbola) and that of the secondary from 2.0 to 2.2389. These changes completely remove both the spherical aberration and the OSC, as can be seen in the fourth ray trace in Table XIV.

The amateur telescope maker finds it almost impossible to make the mirrors required for these well-corrected systems, especially the convex hyperboloid of the classical Cassegrain. He is therefore tempted to use the Dall–Kirkham design, in which the secondary is a convex sphere while

TABLE XIV

TWO-MIRROR TELESCOPE SYSTEMS

c	-0.125	-0.3333333	
d		-3	
n	1	-1	1

		Paraxial ray	
ϕ	0.25	-0.6666666	$f' = 12.0$
$-d/n$		-3	
y	1	0.25	$l' = 3.0$
nu	0	0.25	0.0833333

		Spherical surfaces	
Q	1.0	0.2401960	$F' = 11.522999$
Q'	0.9843135	0.2435727	$L' = 2.806690$
U	0	-14.36151	4.97856
Y	1.0	0.2453707	$LA' = -0.19331$
X	-0.0627461	-0.0100513	$OSC = 0.009013$

		Classical Cassegrain	
e	*1.0*	*2.0*	
Q	1.0	0.2461538	$F' = 12.020833$
Q'	0.9846154	0.2495667	$L' = 3.0$
U	0	-14.25003	4.77189
Y	1.0	0.251309	$LA' = 0$
X	-0.0625	-0.0104710	$OSC = 0.001736$

		Ritchey–Chrétien	
e	*1.0368*	*2.2389*	
Q	1.0	0.2465948	$F' = 12.0000$
Q'	0.9846376	0.2500017	$L' = 3.0000$
U	0	-14.24178	4.78019
Y	1.0	0.2517515	$LA' = 0$
X	-0.062482	-0.0104896	$OSC = 0$

		Dall–Kirkham	
e	*0.839926*	*0*	
Q	1.0	0.2444135	$F' = 12.104064$
Q'	0.9845275	0.2478498	$L' = 3.0000$
U	0	-14.28260	4.73900
Y	1.0	0.2495620	$LA' = 0$
X	-0.0625721	-0.0103982	$OSC = 0.008672$

the primary is a concave ellipse. A few trials reveal the desired eccentricity of
this ellipse in any particular case. For our example the primary ellipse
should have an eccentricity of 0.839926, as shown in the fifth ray trace in
Table XIV. It is clear that the real problem here is coma, which is five times as

TABLE XV

A CLASSICAL CASSEGRAIN SYSTEM

		Concave		Convex		
c		-0.1		-0.4		
d			-4			
n	1		-1		1	
ϕ		0.2		-0.8		
$-d/n$			-4			
y		1		0.2		$l' = 5.0$
nu	0		0.2		0.04	$f' = 25.0$

large as in the classical Cassegrain. Obviously it is wrong to strengthen the rim of the primary, as in the Dall–Kirkham, when it should be weakened, as in the Ritchey–Chrétien form. However, the Dall–Kirkham does have the additional advantage that the elliptical primary can be tested in the workshop before assembly by the use of a pinhole source at one focus and a knife-edge at the other. In our example the two focal lengths are 4.35 and 50.0, respectively.

B. A MAKSUTOV CASSEGRAIN SYSTEM

Many Cassegrain systems have been constructed using only spherical mirrors, the spherical aberration being corrected by means of a meniscus corrector lens placed in the entering beam. The secondary mirror can be conveniently formed by depositing an aluminized reflecting disk on the rear surface of the corrector.

As an example, suppose that before adding the corrector lens we have two mirrors separated by a distance of 4.0, the concave primary placing its image at 1.0 units behind the secondary mirror, which in turn projects the final image to a point 1.0 behind the primary mirror, through a hole. The focal length of the primary is 5.0, and the secondary mirror magnifies this by $5\times$ giving an overall focal length of 25. The paraxial ray trace is shown in Table XV. It will be seen that the l' after the concave mirror is -5.0, and this value must be maintained after adding a corrector lens in order for the final image to remain at a distance of 1.0 behind the primary mirror. This is achieved by recalculating the curvature of the primary mirror each time that a change is made in the system.

To design the correcting lens, we start with some guessed value of c_1, retain the radius of the secondary mirror $c_2 = c_4 = -0.4$, trace a paraxial ray to solve for c_3 to give the desired back focus, and then add a marginal ray at $f/10$. This gives us the spherical aberration and also the $(D - d) \Delta n$ value arising at the lens. After tracing a paraxial principal ray through the front lens vertex, we find l'_{pr} and so determine the OSC. Our trials ran as follows:

	c_1	c_3	LA'	f'
	-0.5	-0.117763	$+14.7$	20.5666
	-0.42	-0.101396	-0.4475	23.8626
	-0.43	-0.103723	$+0.5673$	23.3938
(Setup A)	-0.425	-0.102571	$+0.0397$	23.6259

Taking this last case for further study, we find that the zonal aberration is -0.0170, the $(D - d) \Delta n$ value is -0.0000009 (insignificant), and $l'_{pr} = -1.204$, giving $OSC = -0.00226$.

To investigate the seriousness of this OSC residual we next make a meridional ray plot at an angular field of, say, $-1.5°$ (Fig. 193). This is a

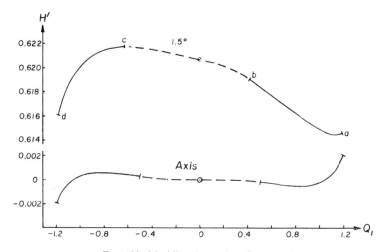

FIG. 193. Meridional ray plot of system A.

very small field angle, but it serves to determine the size of the hole in the primary mirror, and if the field is too wide there will be very little mirror left for image formation. The paths of the upper and lower limiting rays are shown in Fig. 194, where it will be seen that the hole in the primary mirror is the determining factor as to which rays get through and which do not. A front view of the system, looking upwards along the $1\frac{1}{2}°$ beam is shown in Fig. 195.

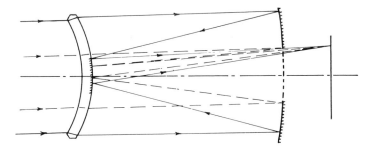

FIG. 194. Ray diagram of setup *A*.

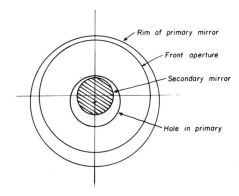

FIG. 195. Front view of system, looking upwards at 1.5° obliquity.

There are two branches to the meridional ray plot, the left-hand branch containing those rays that strike the primary mirror below the hole and the right-hand branch containing those above the hole. There is obviously a large amount of negative coma in this system and there is some degree of inward-curving field, although the Coddington fields are meaningless here since the principal ray is blocked out by the secondary mirror. The Petzval sum, arising mainly at the secondary mirror, is very large (0.5863).

The most effective way to improve this system is to increase the central air space. We will therefore increase this to 5.0, and to maintain the focal length at 25.0 we repeat the paraxial layout (Table XVI).

After adding the corrector lens, we must determine the curvature of the primary mirror, at such a value that the $l'_3 = -6.978947$ in order to place the image once more at 1.0 behind the hole in the primary mirror. Following

TABLE XVI

A CASSEGRAIN WITH INCREASED SEPARATION

c	-0.076		-0.233333		
d		-5			
n	1	-1		1	
ϕ	0.152		-0.466666		
$-d/n$		-5			
y	1		0.24		$l' = 6.0$
nu	0	0.152		0.04	$f' = 25.0$

the previous procedure we end up with the following system (setup B):

	c	d	n_c	Glass
	-0.249			
		0.25	1.52111	K-4
	-0.233333			
		5.0		
(concave)	-0.0786549			
		5.0		
(convex)	-0.233333			

with $f' = 23.82816$, $l' = 6.0000$, $LA' = 0.00272$, $LZA = -0.00383$, $OSC = 0.00014$, $Ptz = 0.3046$.

The meridional ray plot is shown in Fig. 196, and the lens will be seen to be almost perfect except for a strongly inward-curving field. To remove the Petzval sum entirely requires that the two mirrors have the same radius; this

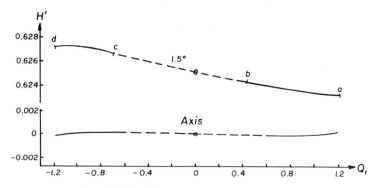

FIG. 196. Meridional ray plot of longer system B.

occurs when the central air space is about 9.0. At an even longer space, at 12.0, the secondary mirror becomes a plane and all the power is in the primary. At these increased lengths the coma correction becomes a problem.

In an effort to reduce the Petzval sum with a reasonably short system, we can insert a negative field flattener in the hole in the primary mirror. We must then redetermine the radius of the primary to restore the back focus at 1.0 beyond the field flattener. The addition of this negative lens increases the focal length (the system is now an extreme telephoto), and it makes both the spherical aberration and the OSC more positive. The spherical aberration can be corrected by an adjustment of c_1, giving setup C:

	c	d	n
Corrector lens	$\left\lvert\begin{array}{l} -0.24865 \\ -0.233333 \end{array}\right.$	0.25	1.52111
		5.0	
Primary mirror	-0.0786009		
		5.0	
Secondary mirror	-0.233333		
		5.13	
Field flattener	$\left\lvert\begin{array}{l} -0.5 \\ 0 \end{array}\right.$	0.1	1.52111

with $f' = 30.325$, $l' = 0.099991$, $LA' = 0.01809$, $LZA = -0.01401$, $OSC = 0.00090$, $\mathrm{Ptz} = 0.1329$.

The 1.5° meridional ray plot of this lens is shown in Fig. 197 alongside

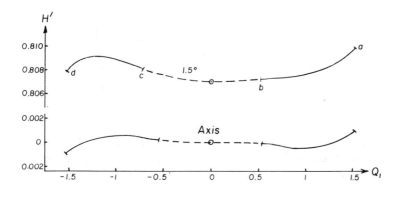

FIG. 197. Meridional ray plot of system C.

that of the axial image. The zonal aberration has become much larger and the positive coma is decidedly serious. The coma can be reduced by slightly reducing the central air space, as in the following system D:

c	d	n
$-0.28192 \,\big\rbrace$ $-0.265446 \,\big\rbrace$	0.25	1.52111
	4.74	
-0.0837446		
	4.74	
-0.265446		
	4.84	
$-0.5 \,\big\rbrace$ $0 \,\big\rbrace$	0.1	1.52111

with $f' = 30.0756$, $l' = 1.00001$, $LA' = 0.03100$, $LZA = -0.02070$, $OSC = 0.00047$, $Ptz = 0.1865$. We have evidently not gone quite far enough since the OSC is still positive. This final system is illustrated in Fig. 198 and

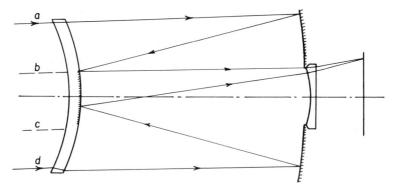

FIG. 198. Ray diagram of final setup D. [For the sake of clarity the limiting rays (b) and (c) have not been drawn in.]

its meridional ray plot at 1.5° in Fig. 199. It will be seen that shortening the system has indeed reduced the coma, but it has greatly increased the zonal spherical aberration, which causes the ends of the 1.5° graph to depart quickly from the desired form.

Systems of this general type have been used frequently in the longer focal lengths for 35 mm SLR cameras. The field angles can be readily found since the picture diagonal is 43 mm:

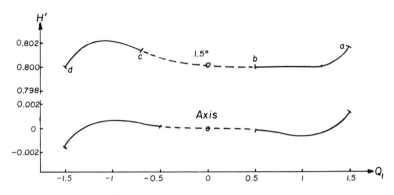

FIG. 199. Meridional ray plot of final system D.

Focal length (mm)	Semifield (deg)
500	2.5
750	1.6
1000	1.2

It must be remembered that these reflecting systems need very careful and complete internal baffling to prevent light from going straight to the film without being reflected by the mirrors. Also it is almost impossible to introduce an iris diaphragm to vary the lens aperture, so all the exposure control must be made by varying the shutter speed.

C. A SCHWARZSCHILD MICROSCOPE OBJECTIVE

It was discovered by Karl Schwarzschild about 1904 that a two-mirror system of the reversed telephoto type, i.e., one in which the entering parallel light first encounters the convex mirror from which it is reflected over to the large concave secondary mirror, has the remarkable property that if both mirrors are spherical and have a common center C, then the primary spherical aberration, coma, and astigmatism are all automatically zero provided the ratio of the mirror radii is equal to $(\sqrt{5} + 1)/(\sqrt{5} - 1) = 2.618034$.[11] This conclusion can be easily verified, for primary aberrations, by use of the surface contribution formulas given on p. 207.

[11] P. Erdös, Mirror anastigmat with two concentric spherical surfaces, *J. Opt. Soc. Am.* **49**, 877 (1959).

At finite aperture this system suffers from a very small spherical overcorrection, an example being as follows:

	c	d
(convex)	1.0	
		1.618034
(concave)	0.381966	

with $f' = 0.809017$, $l' = 3.427051$, LA' $(f/1) = 0.00137$, OSC $(f/1) = 0.00129$, LA' $(f/2) = 0.00008$, OSC $(f/2) = 0.00007$. For the OSC calculation it was assumed that the stop is at the concave mirror, making $l'_{pr} = 0$, whence $OSC = (F'l'/f'L) - 1$.

However, when this system is intended for use as a microscope objective, the object must be at such a finite distance as to give the desired magnification. To correct the spherical aberration and coma it is then necessary to weaken the concave mirror appropriately, the two mirrors remaining concentric about the common center C. The separation is, of course, equal to $r_1 - r_2$. A few trials tell us quickly what separation should be used. The following is an example of a $10\times$ objective of this type:

	c	d
(convex)	1.0	
		2.07787
(concave)	0.3249	

with

$$L = l = -7.14694, \qquad \sin U = u = -0.05$$

$$l' = 3.89256, \qquad m = -0.1$$

$$\text{N.A.} = 0.5: \qquad LA' = -0.000002, \quad OSC = -0.000003$$

$$\text{N.A.} = 0.35: \qquad LA' = -0.000394, \quad OSC = -0.000382$$

There is a small zonal residual of spherical aberration, decidedly less than the zonal tolerance of $6\lambda/\sin^2 U'_m$ given on p. 121, which in this case amounts to 0.00052, assuming that the unit of length is the inch. A scale drawing of the system is shown in Fig. 200. It will be seen that the diameter of the convex mirror must be 0.72 to catch the marginal ray at N.A. = 0.5, and this blocks out the middle of the beam so that the lowest ray has a N.A.

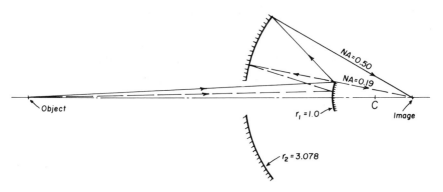

FIG. 200. A Schwarzschild microscope objective.

of 0.193. The diameter of the hole in the concave mirror must be about 0.56, but this is not a limiting aperture and it can be made somewhat larger. However, if it is too large it will pass unwanted light, which is undesirable. The obstruction is not large enough to cause a serious degradation of the definition.

D. THREE-MIRROR SYSTEM

An interesting modification of the classical Gregorian reflecting system has been suggested by Shafer,[12] in which the marginal ray is reflected twice at the primary mirror, at equal distances on opposite sides of the axis, and once at the concave secondary mirror, the final image being formed at the center of the secondary mirror. The paraxial layout is shown in Fig. 201. If

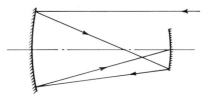

FIG. 201. A three-reflection aplanatic system.

the mirror radii are respectively 10 and 7.5, the separation is 7.5 and the focal length is − 7.5. A few trials indicate that for the simultaneous correction of spherical aberration and OSC with a semiaperture of 1.0 (i.e., an f/3.75

[12] D. R. Shafer, New types of anastigmatic two-mirror telescopes, *J. Opt. Soc. Am.* **66**, 1114, Abs. ThE-17 (1976).

system), the primary must be a concave ellipse with eccentricity 0.63782 and the secondary a concave hyperboloid with eccentricity 2.44. The aperture stop and both the pupils are at the primary mirror. Since the middle half of the entering beam is blocked out by the secondary mirror, the image receiver can cover the middle half of the secondary mirror without introducing any further obstruction. The angular field of our $f/3.75$ system is therefore $\pm 1.91°$. If the aperture is doubled, the angular field will also be doubled to $\pm 3.81°$.

CHAPTER 15

Eyepiece Design

An eyepiece differs fundamentally from a photographic objective in that the entrance and exit pupils are outside the system. The lens itself must therefore have a large diameter, which is determined far more by the angular field to be covered than by the relative aperture. The latter is set by the objective lens and has little relation to the structure of the eyepiece itself.

So far as aberration correction is concerned, the axial spherical and chromatic are usually unimportant, and they can be corrected in the objective if necessary. Lateral color and coma must be corrected as well as possible. Most eyepieces have a large Petzval sum, which leads to a large amount of astigmatism at the edge of the field. Because the observer naturally prefers to relax his accommodation on axis and accommodate as much as necessary when viewing at the edges of the field, it is customary to aim at a flat sagittal field and a backward-curving tangential field, including the objective, relay (if any), prisms, etc. in the computation. An attempt to reduce the astigmatism by making the tangential field less backward-curving generally leads to an inward-curving sagittal field, which is unpleasant to the observer. Of course, the situation is much improved if some way can be found to reduce the Petzval sum of the entire system, but this is difficult because the eyepiece has a short focal length, and therefore a large Petzval sum, while the objective has a longer focal length and a smaller Petzval sum.

For easy use, the eyepoint, where the emerging principal ray crosses the axis, should be at least 20 mm from the last lens surface. This is difficult to achieve in a high-power eyepiece, and often requires a deep concavity close to the internal image plane. There may also be a serious amount of spherical aberration of the exit pupil. This causes the principal rays of oblique beams to cross the axis at points that become progressively closer to the rear lens surface at increasing obliquities, so that the eye must be moved forward to view the edges of the field. The eye is then not in the best position to view the intermediate parts of the field, resulting in a "kidney-bean" shadow, which moves about as the eye is moved. One way to correct this is to include a parabolic surface somewhere in the eyepiece, or to insert an aspheric plate in the focal plane to make the extreme principal rays diverge before entering the eyepiece.

All eyepieces suffer from some degree of distortion, of the pincushion type as seen by the eye. This is often of such an amount that the oblique magnifying power is given more by the ratio of the emerging and entering angles themselves than by the ratio of their tangents. When this is the case the distortion amounts to about 6% at an apparent field of 24° and perhaps 10% at 35°. However, because of the circular shape of the eyepiece field these large distortions are seldom bothersome to the observer.

The design procedures for a number of the simpler types of eyepiece have been described by Conrady. These include the Huygenian (p. 484), the Ramsden (p. 497), the Kellner, or achromatized Ramsden (p. 503), the simple achromatic (p. 761), and the various cemented or air spaced triplet types (p. 768). Other more complicated eyepieces have been described by Rosin.[1] In this chapter we will discuss the design of an eyepiece of the excellent so-called military type, consisting of two cemented doublets mounted close together, and also one of the Erfle type commonly used in wide-angle binoculars.

I. DESIGN OF A MILITARY-TYPE EYEPIECE

As an example of the design of an eyepiece of this type, we will assume a focal length of 1 in. and a clear aperture of just over 1 in., for use with a 10-in. telescope doublet objective having a clear aperture of 2 in. ($f/5$). The true field will be $-2.4°$ at the objective, giving an apparent field at the eye of about 25°. It should be noted that in the absence of distortion the apparent field would be given by the tangent ratio being equal to the focal-length ratio, or $\tan U'_{pr} = 10 \tan 2.4°$, whence $U'_{pr} = 22.7°$. The actual emerging ray slope is more likely to be about 25°, with about 10% distortion.

A. THE OBJECTIVE LENS

For the objective lens, we will take the $f/5$ aplanatic doublet described on p. 173, scaled down to $f' = 10.0$ in e light:

	c	d	n_e	
	0.1554			
		0.32	1.56606	(SK-11)
	−0.2313			
		0.15	1.67158	(SF-19)
$(D-d)$	−0.0549164			

---

[1] S. Rosin, Eyepieces and magnifiers, in "Applied Optics and Optical Engineering" (R. Kingslake, ed.), Vol. III, p. 331. Academic Press, New York, 1965.

with $f' = 9.99963$, $l' = 9.76247$, $LA'(f/5) = 0.00048$, $LZA(f/7) = -0.00168$, $OSC(f/5) = 0.00011$. The upper and lower rim rays at $-2.4°$ are next traced with $Y_1 = \pm 1.0$ at the front vertex, and also the principal ray midway between them. By Coddington's equations traced along the principal ray we find

$$H'_{pr} = 0.419107, \qquad X'_s = -0.01455, \qquad X'_t = -0.03154$$

B. EYEPIECE LAYOUT

To lay out the eyepiece, we may decide to use the same glasses as for the objective, and keep the outside surfaces plane. As a start we can make all the other surfaces of the same curvature, which for the prescribed focal length of 1.0 is 1.0337. Tentative thicknesses are set at 0.4 for the crowns and 0.1 for the flints with a separation of 0.05. As a check on these thicknesses we trace the lower rim ray entering the objective at $-2.4°$, and we find that it intersects the six surfaces of the eyepiece at these heights:

Field lens	Eye lens
0.5175	0.4877
0.5400	0.4362
0.5490	0.3854

A scale drawing (Fig. 202) indicates that the thicknesses of the crown ele-

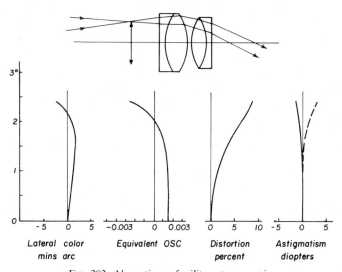

FIG. 202. Aberrations of military-type eyepiece.

ments should be changed to 0.5 and 0.35, respectively. Having done this, we restore the focal length by changing c_4 to 1.0237.

1. *Lateral Color*

Our first task is to calculate the angular lateral color $U'_F - U'_C$ at the eye by tracing principal rays in C and F light through the entire system including the objective lens. It is best to do this at two obliquities so that a nice balance can be obtained. This gives the following:

Field angle (deg)	$U'_F - U'_C$ (deg)	Minutes of arc
2.4	−0.0505	−3.03
1.5	+0.0241	+1.44

This might well be considered an excellent balance since it favors the intermediate parts of the field and lets the extreme edge go. However, we can slightly reduce the lateral color at 2.4° by weakening c_5 to −1.0 and holding the focal length of the eyepiece by changing c_4 to 1.0226. This brings the lateral color at 2.4° to −2.30 minutes of arc, which we will accept.

2. *Coma*

We must now direct attention to the coma. This is found by tracing the upper and lower rim rays through the whole system, and then calculating their point of intersection by the formulas given on p. 143. The vertical distance of this point above or below the principal ray is a direct measure of the tangential coma; we get the "equivalent OSC" by dividing the coma$_t$ by 3 and by H'_{pr}. Once again it is best to calculate this at two obliquities and try to secure the best balance between them, letting the extreme value go somewhat in order to favor the intermediate fields. Our present system gave the following:

Field angle (deg)	L'_{ab}	H'_{ab}	H'_{pr}	Coma$_t$	Equivalent OSC
2.4	−9.6221	4.6920	4.7415	−0.0496	−0.00348
1.5	−147.927	40.2579	40.3045	−0.0466	−0.00039

We must obviously try to make the coma more positive. We can do this by weakening the field lens, say by 5%, and then repeating the correction of the lateral color and focal length. These changes gave the following system:

	c	d	n_e
Field lens	0		
		0.1	1.67158
	0.982		
		0.5	1.56606
	−0.982		
		0.05	
Eye lens	1.07227		
		0.35	1.56606
	−1.03		
		0.1	1.67158
	0		

with $f' = 1.0000$, $l = -0.59779$; lateral color: $2.4° = -2.43$ min, $1.5° = +1.79$ min; distortion: $2.4° = 8.23\%$, $1.5° = 3.13\%$. For the coma we find the following:

Field angle (deg)	U'_{pr} (deg)	L'_{ab}	H'_{ab}	H'_{pr}	$Coma_t$	Equivalent OSC
2.4	24.4	−9.7363	4.6896	4.7282	−0.0385	−0.00271
1.5	15.1	−410.79	111.275	111.1386	+0.1263	+0.00114
					Paraxial:	+0.00156

The paraxial *OSC* is assumed to be equal and opposite to the *OSC* at the internal image, found by tracing a marginal ray back into the eyepiece from the exit-pupil. Since these corrections appear to be reasonable, we next turn to the astigmatism.

3. *Astigmatism*

The astigmatism of the system is found by calculating Coddington's equations along the traced principal rays, including the objective lens as well as the eyepiece. The closing formulas give the oblique distances s' and t' from the eyepoint, which is here assumed to be at a distance of 0.7 beyond the rear surface. It is more meaningful to convert the final s' and t' values to diopters of accommodation at the eye; this is done by dividing the calculated values into 39.37, the number of inches in a meter, and reversing the sign. For our last system we have the following:

Field angle (deg)	s'	t'	Diopters at eye s'	t'	Eye relief L'_{pr} (in.)
2.4	34.054	−12.076	−1.16	+3.26	0.69
1.5	62.94	−136.19	−0.63	+0.29	0.76
				Paraxial:	0.81

In this table, a positive diopter value represents a backward-curving field that the observer can readily accommodate; a negative sign indicates an inward field, which requires the observer to accommodate beyond infinity, an almost impossible requirement for most people. Thus the negative values should be kept as small as possible, and certainly less than one diopter. The various aberrations of this final system are shown graphically in Fig. 202.

II. AN ERFLE EYEPIECE

When it is desired to provide an apparent angular field approaching $\pm 35°$, it is necessary to weaken the inner convex surfaces of the two-doublet "military" eyepiece and insert a biconvex element between them. This type of eyepiece was patented in 1921 by H. Erfle.[2]

Because of the great length of the eyepiece, and because the clear aperture must be considerably greater than the focal length, it is usual to weaken the field lens and provide a deep concave surface close to the internal image plane, so as to keep the eye relief as long as possible. The concave surface near the image also helps reduce the Petzval sum (Fig. 203).

In view of these considerations, we will assign a power of 0.1 to the field lens, 0.4 to the middle lens, and the eyelens will then come out to have a power of about 0.36 for an overall focal length of 1.0. This is an entirely arbitrary division of power and some other distribution might be better. We will use the same glasses as for the military eyepiece, with BK-7 for the middle lens. Since we have more degrees of freedom than we need to correct three aberrations, we can make some of the positive elements equiconvex for economy in manufacture. The starting system, to be used with the same objective lens as before, will be as follows:

---

[2] H. Erfle, U.S. Patent 1,478,704, filed Aug. 1921.

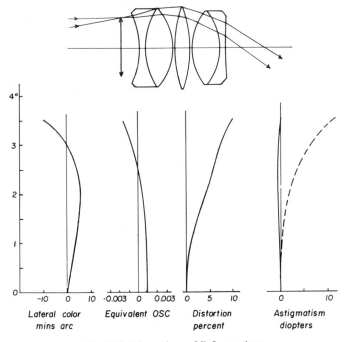

FIG. 203. Aberrations of Erfle eyepiece.

	c	d	n_e
Field lens $\phi = 0.1$	$\begin{cases} -0.6 \\ \\ 0.6 \\ \\ -0.833563 \end{cases}$	0.1 0.6	1.67158 1.56606
Middle lens $\phi = 0.4$	$\begin{cases} 0.3949 \\ \\ -0.3949 \end{cases}$	0.05 0.35 0.05	 1.51871
Eye lens	$\begin{cases} 0.8175 \\ \\ -0.8175 \\ \\ 0.05 \end{cases}$	0.6 0.1	1.56606 1.67158

with $f' = 1.0$, $l = -0.34460$, $2.5°$ lateral color $= 8.38$ min arc.

Clearly, our first act must be to reduce the lateral color; to do this we strengthen c_7 and solve for the overall focal length by c_6. The chosen thicknesses are just sufficient to clear the 3.5° beam from the objective. Our second setup is as follows:

c	d	n_c
-0.6		
	0.1	1.67158
0.6		
	0.6	1.56606
-0.833563		
	0.05	
0.3949		
	0.35	1.51871
-0.3949		
	0.05	
0.83321		
	0.6	1.56606
-0.96		
	0.1	1.67158
0.05		

with $f' = 1.0$, $l = -0.34987$; lateral color: $3.5° = -5.67$ min, $2.5° = +5.58$ min; equivalent OSC: $3.5° = -0.00301$, $2.5° = -0.00049$, $1.5° = 0.00057$, axis $= -0.00096$. This lateral color is probably satisfactory, although an increase in the negative value at 3.5° would be advantageous since it would tend to reduce the lateral color at the intermediate fields. As before, the so-called equivalent OSC was found by tracing upper, principal, and lower rays at each obliquity and finding the intersection of the upper and lower rays in relation to the principal ray height. The coma$_t$ so found was divided by $3H'$ as before to give the equivalent OSC. For the axial OSC, a marginal ray was traced backwards, entering the eye-lens parallel to the axis at a height of $Y_1 = 0.1$, and finding the ordinary OSC at the internal image. The equivalent OSC at the eye was then taken as being equal and opposite to the true OSC at the internal image.

It is clear that we must reduce the negative OSC at the 3.5° obliquity. The simplest way to do this is to strengthen the interface c_2 in the field lens and readjust the interface in the eye-lens to restore the lateral color correction, always holding the focal length by c_6. It was also found advantageous to deepen c_8 slightly and to reduce the two air spaces between the elements. With all these changes we get the following:

		c	d	n_e
Field lens $\phi = 0.1$	-0.6			
			0.1	1.67158
	0.7			
			0.6	1.56606
	-0.846516			
			0.03	
Middle lens $\phi = 0.4$	0.3949			
			0.35	1.51871
	-0.3949			
			0.03	
	0.83941			
			0.6	1.56606
Eye lens	-0.85			
			0.1	1.67158
	0.1			

with $f' = 1.0$, $l = -0.37806$; at internal image: $LA = +0.00612$ (under-correction), $OSC = -0.00099$ (overcorrection). Then

Field (deg)	U'_{pr} (deg)	Lateral color (min)	L'_{ab}	Equivalent OSC	Distortion (%)	Diopters		
						s'	t'	L'_{pr}
3.5	33.9	-9.57	-2.32	-0.00150	9.50	-0.09	$+11.04$	0.57
2.5	24.7	$+4.90$	-8.93	-0.00012	5.75	-0.51	$+3.88$	0.64
1.5	14.9	$+5.78$	-54.20	$+0.00068$	2.17	-0.27	$+0.91$	0.69
			Paraxial:	$+0.00099$			Paraxial:	0.72

The properties of this eyepiece are shown graphically in Fig. 203. There is a good balance in the lateral color and also in the equivalent OSC. The tangential field is decidedly backward-curving, which is desirable, especially since the sagittal field is flat. The only sure way to change the field curvature is to redesign the entire eyepiece with other glasses, chosen to have a smaller index difference across the internal surfaces, but keeping a large V difference for the sake of lateral color correction.

III. A GALILEAN VIEWFINDER

The common eye-level viewfinder used on many cameras is a reversed Galilean telescope, with a large negative lens in front and a small positive lens near the eye. The rim of the front lens serves as a mask to delimit the viewfinder field, but of course since it is not in the plane of the internal image, there will be some mask parallax and the mask will appear to shift relative to the image if the observer should happen to move his eye sideways.

To design such a viewfinder, it is necessary to specify the size of the negative lens, the length of the finder, and the angular field to be covered in the object space. It is usual to assume that the eye will be located about 20 mm behind the eye lens. The magnifying power of the system follows from the given dimensions. The axial magnifying power is given by the ratio of the focal length of the negative lens to the focal length of the eyelens, which is the ratio y_1/y_4 for a paraxial ray entering and leaving parallel to the lens axis. The oblique magnifying power is given by $\tan U'_{pr}/\tan U_{pr}$ and generally varies across the field. It can be made equal to the axial magnifying power, to eliminate distortion, by the use of an aspheric surface on the rear of the front lens; a concave ellipsoid is a useful form for this aspheric.

As an example, we will design a Galilean viewfinder having a front negative lens about 30 mm diagonal to cover a $\pm 24°$ field, a central lens separation of 40 mm, and an eyepoint distance of about 20 mm. We start by guessing at a possible front negative element. A paraxial ray is traced through it, entering parallel to the axis, and by a few trials we ascertain the radii of a small equiconvex eyelens to make the system afocal. A 24° principal ray is then traced with a starting Q_1 equal to 15 mm, and the oblique magnifying power and L'_{pr} are found. The distortion is also calculated by $MP_{oblique} - MP_{axial}$.

A concave ellipse is then substituted for the second spherical surface, of course, with the same vertex curvature so as not to upset the paraxial ray, and by experimentally varying its eccentricity the distortion can be eliminated. If the L'_{pr} is then about 20 mm the problem is solved. If not, then it is necessary to change c_2 and repeat the whole process.

The following design resulted from the procedure just outlined (all dimensions in centimeters):

	c	d	n
Ellipse with	0.1		
$e = 0.5916$		0.30	1.523
$(1 - e^2) = 0.65$	0.38		
		4.00	(air)
	0.089698		
		0.25	1.523
	−0.089698		

with $L'_{pr} = 2.043$; magnifying power: $24° = 0.6250$, $15.8° = 0.6247$, axis $= 0.6249$; focal length: front lens $= -6.686$, rear lens $= 10.699$. After tracing the corner principal ray at 24°, to locate the eyepoint, other principal rays can be traced right-to-left through this eyepoint out into the object space. It will be seen that this particular elliptical surface has completely eliminated the distortion. A diagram of the system is given in Fig. 204. In practice, of

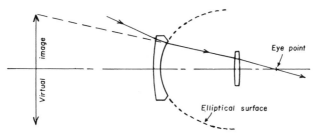

FIG. 204. A Galilean eye-level viewfinder.

course, the front lens is cut into a square or rectangular shape to match the format of the camera, and to match its vertical and horizontal angular fields. For safety, the viewfinder is often constructed to indicate a field slightly narrower than that of the camera itself.

CHAPTER 16

Automatic Lens Improvement Programs[1]

The methods of lens design outlined in this book were the only procedures available up to about 1956, when electronic computers became available having sufficient speed to be used for lens design. Many people in several countries then began work on the problem of how to use a high-speed computer, not only to trace sufficient rays to evaluate a system but to make changes in the system so as to improve the image quality. It is our purpose in this chapter to indicate how such a computer program is organized and how the various "boundary conditions" are handled.[2]

When using this type of program, a starting system is entered into the program, and the computer then proceeds to make changes that will reduce a calculated "merit function" to its lowest possible value. The starting system need not be a particularly good lens, and often a very rough approximation to the desired system can be used. Indeed, some designers have even submitted a set of parallel glass plates to the computer, leaving it up to the program to introduce curved surfaces where necessary. Lenses designed in this way are not likely to be as good as those in which the initial starting system is already fairly well corrected.

I. CASE OF AS MANY ABERRATIONS AS THERE ARE DEGREES OF FREEDOM

We will consider first the simple case of a lens having the same number of degrees of freedom, N, as there are aberrations to be corrected. By degrees of freedom, or variables, in a lens we refer to the surface curvatures, air spaces, and sometimes lens thicknesses, although thickness changes do not generally help very much.

We first evaluate all the aberrations of our starting system. We then make small experimental changes in each of the N variables in turn, and evaluate the change in each aberration resulting from this small change in

---

[1] The author acknowledges with thanks the help of Donald P. Feder in writing this chapter.

[2] D. P. Feder, Automatic optical design, *Appl. Opt.* **2**, 1209 (1963).

each variable. (This procedure was followed in the design of a telephoto lens on p. 264.)

To remove all the aberrations we must now solve N equations of the form

$$\Delta ab_1 = \left(\frac{\partial ab_1}{\partial v_1}\right)\Delta v_1 + \left(\frac{\partial ab_1}{\partial v_2}\right)\Delta v_2 + \left(\frac{\partial ab_1}{\partial v_3}\right)\Delta v_3 + \cdots$$

$$\Delta ab_2 = \left(\frac{\partial ab_2}{\partial v_1}\right)\Delta v_1 + \left(\frac{\partial ab_2}{\partial v_2}\right)\Delta v_2 + \left(\frac{\partial ab_2}{\partial v_3}\right)\Delta v_3 + \cdots$$

$$\vdots$$

where ab represents an aberration and v represents a variable, or degree of freedom, in the lens. There are, of course, N equations here in N unknowns, and there are N^2 coefficients that we must evaluate by making small experimental changes.

Provided the variables have been chosen to be effective in changing the particular aberrations, so that the equations are well conditioned, then the N equations can be solved simultaneously, and, if everything were linear, the solution would tell us how much each variable should be changed to yield the desired changes in the aberrations. Unfortunately, a lens is about as nonlinear as anything in physics, and it will probably happen that some at least of the calculated changes are far too large to be used. So we take a fraction of the changes, say 20–40%, and apply these to the lens parameters. This should yield an improved system, but nowhere near the desired solution. Then we repeat the process, and we must now reevaluate the N^2 coefficients because the changes that we have introduced will alter the course of all the traced rays and hence all the subsequent coefficients. In the next iteration we shall be closer to the solution and the changes will be smaller, so we can take a much larger fraction, say 50–80%. After a third iteration we should be so close to the solution that the whole of the calculated changes can be applied.

II. CASE OF MORE ABERRATIONS THAN FREE VARIABLES

Suppose we have M aberrations and only N variables, where M is greater than N. Then our procedure will yield M equations in N unknowns, and a unique solution is impossible. The equations to be solved can be written in simple form:

$$y_1 = a_1 x_1 + a_2 x_2 + a_3 x_3 + \cdots$$

$$y_2 = b_1 x_1 + b_2 x_2 + b_3 x_3 + \cdots$$
$$\vdots$$

where the y are the desired changes in the aberrations and the x are the changes in the variables. The quantities a, b, ... are the coefficients determined by making small experimental changes in the variables.

Although an exact solution is now impossible, we can ascertain a set of changes x that will minimize the sum of the squares of the aberration residuals R, where

$$R_1 = a_1 x_1 + a_2 x_2 + a_3 x_3 + \cdots - y_1$$
$$R_2 = b_1 x_1 + b_2 x_2 + b_3 x_3 + \cdots - y_2$$
$$\vdots$$

Obviously the R are in the nature of aberrations. Our problem is to find the set of x values that will minimize the sum

$$\phi = R_1^2 + R_2^2 + R_3^2 + \cdots$$

there being as many R as there are aberrations and as many x as there are variables. The sum ϕ is called a merit function, and our aim is to reduce its value as far as possible.

There are two reasons why we sum the squares of the residuals instead of the residuals themselves. One is because all squares are positive, and of course we do not wish to have one negative aberration compensating some other positive aberration. Another reason is that any large residual will be greatly increased on squaring so that it will receive most of the correcting effort of the program, while a small residual when squared becomes smaller still and is ignored by the program. Eventually all the residuals end up at about the same value, and the image point will then be as small as it can become.

A. What Is an "Aberration"?

Of course, if we wish, we can use the conventional aberrations in this solution, provided they are all expressed in some comparable terms such as their transverse measure. However, it would then be difficult to ensure that there will be more aberrations than variables. It is much better to abandon the conventional aberrations and regard as an aberration the departure of each ray from its desired location in the image plane. Thus if we trace 20 rays to evaluate our lens we shall end up with 20 aberrations, already more than the number of variables in most lenses. It is convenient to regard the desired location of each ray as the point where the principal ray of the beam strikes the image plane; we must then regard the distortion of the principal ray as a separate aberration.

For the rays to be traced, a common selection is to take two axial rays, marginal and zonal, with five meridional and two sagittal rays at each of two obliquities, making 16 rays altogether. Of course, we must not forget chromatic aberration, and we may wish to trace a few colored rays in addition. However, many workers prefer to calculate Conrady's $D - d$ sum along the marginal ray for axial achromatism, along the zonal ray for spherochromatism, and along all the traced meridional rays to eliminate the longitudinal and lateral chromatism off-axis. This now adds up to 26 aberrations plus 2 distortions, or 28 altogether. We thus have to determine 28 coefficients for each lens variable, and solve a bank of 28 equations to calculate the changes to be made in these variables.

We have indicated that these coefficients can be found by making small trial changes in the variables and tracing all the rays through each modified system. Many workers, however, have found ways by which the coefficients can be determined by algebraic manipulation of the ray data, which saves a great amount of computing time.[3]

B. SOLUTION OF THE EQUATIONS

For the merit function ϕ to be a mathematical minimum, we must solve a set of equations of the form

$$\frac{\partial \phi}{\partial x_1} = 0, \qquad \frac{\partial \phi}{\partial x_2} = 0, \qquad \frac{\partial \phi}{\partial x_3} = 0, \dots$$

there being as many of these equations as there are variables in the lens. Differentiating our expression for ϕ, we get the appropriate set of equations

$$\frac{\partial \phi}{\partial x_i} = 2R_1\left(\frac{\partial R_1}{\partial x_i}\right) + 2R_2\left(\frac{\partial R_2}{\partial x_i}\right) + 2R_3\left(\frac{\partial R_3}{\partial x_i}\right) + \cdots = 0$$

for $i = 1, 2, \dots, N$.

Entering the successive derivatives of the R with respect to x_1 for the first equation gives

$$\frac{1}{2}\frac{\partial \phi}{\partial x_1} = (a_1 x_1 + a_2 x_2 + a_3 x_3 + \cdots)a_1 + (b_1 x_1 + b_2 x_2 + \cdots)b_1 + \cdots = 0$$

[3] D. P. Feder, Differentiation of raytracing equations with respect to construction parameters of rotationally symmetric optics, *J. Opt. Soc. Am.* **58**, 1494 (1968).

or

$$x_1(a_1^2 + b_1^2 + \cdots) + x_2(a_1 a_2 + b_1 b_2 + \cdots) + \cdots - (a_1 y_1 + b_1 y_2 + \cdots) = 0$$

Carrying out this differentiation in turn for each of the N variables, we obtain the so-called normal equations. They are simultaneous linear equations and have a unique solution. This is the well-known least-squares procedure invented by Legendre in 1805. Programs are available for performing this series of operations on most large computers.

C. Damped Least Squares

In the early stages of the optimization process it is common to find the program demanding large changes in some of the variables, which are then reversed at the next iteration. To prevent this kind of oscillation it was suggested by Levenberg and others that the merit function should be modified to include the sum of the squares of the changes in the variables x so that

$$\phi = \sum R^2 + p \sum x^2$$

The "damping factor" p is made large at first to control the oscillations, but of course when it is large the improvement in the lens is very slow. For each iteration thereafter, the value of p is gradually reduced until the procedure finally becomes an almost perfect least-squares solution with no damping. This process replaces the use of fractions of the calculated changes suggested in Section I.

D. Use of Wave Form instead of Ray Errors

Some programmers prefer to express the aberrations of a lens by an algebraic expression for the shape of the wave front emerging from the lens, instead of calculating the geometric ray displacements in the image plane. These two methods are equivalent, since one of them states the departures of the wave front from the ideal spherical form while the other uses the slope of the wave front, the second being the derivative of the first.

It should be remarked that the use of only a few selected rays ignores possible errors existing over much of the lens aperture, so that some designers, especially toward the end of a design, add a couple of quadrantal rays, one above and one below the sagittal meridian. Some programmers have made use of Coddington's equations traced along selected rays to determine the curvature of the wave front at the point where the ray emerges from the lens.

E. WEIGHTING FACTORS

It is obvious in our selection of rays to be traced that some are more significant than others. For instance, the two axial rays count as much as the seven rays traced at each obliquity, while the rays at a lower obliquity are more important than those near the edge of the field. In some lenses the distortion must be well controlled, while in others it may be completely ignored.

To meet this need we apply a weighting factor to each residual before solving the equations, so that the first equation will appear as

$$R_1 = K_1(a_1 x_1 + a_2 x_2 + a_3 x_3 + \cdots - y_1)$$

where K_1 is the weighting factor to be applied to aberration 1.

F. ASPHERIC SURFACES

If a lens is to have one or more aspheric surfaces, all the terms that define the asphere must be counted as free variables and solved along with the surface curvatures themselves. This greatly increases the complexity of the least-squares solution by adding several more unknowns to the panel of equations.

G. THE APERTURE STOP

In any lens there must be a well-defined aperture stop that limits the diameter of the beams of light traversing the lens. The position of this stop determines which possible rays in an oblique pencil will get through and which will be cut off. This stop must be entered into the program as if it were a lens surface with air on both sides, and its position can be varied by the program as if it were an ordinary refracting surface. The diameter of the stop must, of course, be equal to the diameter of the axial beam at that position in order to maintain the F number of the system.

H. CHANGE OF GLASS

It is obviously a great help to the designer if the computer optimization program is able to make changes in the glasses as well as changing the surface curvatures and the air spaces. However, this is not a simple matter because the glass catalog does not include a continuum of glasses but only a set of separated types and, more important, the range of available dispersions depends on the choice of refractive index. Thus with an index of 1.52, for instance, the Schott catalog contains glasses ranging from 51 to 70 in V number, whereas if the program were to demand a higher index, say 1.62, then the available dispersion range would have to be changed to 35 to 63;

and at an index of 1.72 the V number range is from 29 to 54. Consequently, the programmer must develop algebraic formulas relating the available dispersion range to the refractive index, which will act as a limitation when the program tries to change a glass. Whenever this feature is available in a program, it seems that the crown glasses tend to move up to the northwest part of the glass chart in the region of the most expensive and extreme lanthanum crowns, while the flint glasses move back and forth along the old flint line. However, the designer is always free to fix one or more of the glasses and let the program change the others in its process of optimization.

I. ZOOM LENSES

A zoom lens presents many difficulties to the programmer. The lens must obviously be equally well corrected at several points in the zoom range; therefore the merit function must contain aberration data calculated at each of these configurations. Also, the zoom law that is initially set up by paraxial ray traces through a series of thin elements must be interpreted by the computer in such a way that after inserting suitable thicknesses, the image plane will remain fixed while the focal length is varied. It is necessary for the designer to specify which spaces can be changed to vary the focal length and which can be changed by the program to improve the aberration corrections. It is extremely necessary that the designer provide a sufficient number of lens elements in a zoom system so that the program can correct the aberrations adequately at several zoom positions.

III. CONTROL OF THE BOUNDARY CONDITIONS

In addition to the reduction of the merit function to improve the image quality, a computer optimization program must be able to control several so-called boundary conditions, for otherwise the lens may be unmakable. The principal boundary conditions that must be controlled are as follows.

A. FOCAL LENGTH

It is obvious that in any lens design the focal length must be maintained at a fixed value within a small tolerance, for otherwise the relative aperture, the angular field, and everything else will change as the design proceeds.

One way to hold the focal length is to solve the last radius to give the desired value of u'. However, this is not a good procedure since it leads to a transfer of power from various parts of the system to the rear element, which may not be at all good for the other corrections. Another method is to make all the changes indicated by the least-squares solution and then ascertain what the focal length has become; the entire system is then scaled up or down to restore the focal length before the next iteration. This is better, but it may lead to a very large lens with perhaps unreasonable thicknesses. The best method is to regard an error in focal length as an aberration, with a fairly heavy weight if it is important to hold the focal length precisely.

The back focus, or image distance from the rear of the lens, can be handled in the same way. Now there is no objection to its being too long but there is a very serious objection if it is too short or even negative. We thus assign a high weight to the value of $l'_{found} - l'_{desired}$ if it is negative, and a zero weight if it is positive.

There is a very real problem here known as boundary bounce. If the program calls for a value of l' outside the tolerance to correct the aberrations, the use of a high weight on the back focus will force the lens into an undesirable configuration, which the program will constantly try to improve. Thus every alternate iteration will tend to push the back focus out of tolerance, and the next iteration will force it back into the tolerable region again. This oscillation takes up a great deal of computer time and makes it hard for the program to yield a well-corrected system.

B. THE LENGTH OF A LENS

It is commonly found that lens-improvement programs tend to make a lens get progressively longer with successive iterations. To prevent this, it is good to regard the length of the lens between the end vertices as an aberration, to be included in the merit function in the ordinary way. This is called an elastic constraint, and if the weighting factor is suitably chosen it can be very effective in holding the lens to a reasonable size.

C. INTERSECTION OF ADJACENT SURFACES

To eliminate the possibility that two adjacent lens surfaces intersect, it is customary to calculate the length along every traced ray from each surface to the next. So long as these distances are greater than some assigned tolerance no action need be taken, but if this length ever drops below the tolerance, the axial separation between the surfaces must be increased to remove the violation before the rays are allowed to continue on their way.

A more sophisticated approach to this problem is to link the variables together by subsidiary equations that prevent boundary violations from occurring. If, for example, the separation t between adjacent surfaces may never be less than t_0, clearly t is a function of the changes x in the lens parameters, so that

$$t - t_0 = \left(\frac{\partial t}{\partial x_1}\right)x_1 + \left(\frac{\partial t}{\partial x_2}\right)x_2 + \left(\frac{\partial t}{\partial x_3}\right)x_3 + \cdots$$

When t reaches the limit t_0 the right-hand side of this equation becomes zero. If $\partial t/\partial x_1$ is not zero, we can then solve for x_1 by

$$x_1 = -\left[\left(\frac{\partial t}{\partial x_2}\right)x_2 + \left(\frac{\partial t}{\partial x_3}\right)x_3 + \cdots\right]\bigg/\left(\frac{\partial t}{\partial x_1}\right)$$

We now substitute this value of x_1 in all the aberration equations, thus reducing the number of equations by one. This "reduced" set of equations is now solved for the remaining variables x_2, x_3, etc., by the least-squares procedure. Finally x_1 is found by direct substitution into the above formula. Although it has been indicated that x_1 should be selected in this way, any of the variables could have been used provided its derivative were not so small as to produce rounding errors. This procedure can be used to control surface separations, back focus, and focal length variations.

D. VIGNETTING

In very few lenses do the oblique beams completely fill the aperture stop. It is far more common for the designer to permit some degree of vignetting to exist, to keep the lens diameters reasonably small and also to cut off extreme rays that behave badly and would cause a loss of definition if allowed to pass. If the designer specifies, for instance, that the 20° beam shall have a vertical dimension equal to 75% of the height of the axial beam, then the upper and lower limiting rays of the 20° beam must be carefully chosen so that the upper ray crosses the stop at 0.75 of the height of the stop above the axis, and the lower ray crosses the stop at 0.75 of the stop height below the axis. It is therefore necessary to make a few exploratory ray traces to discover the entering data of the desired upper and lower rays before any aberrations are calculated. It is unlikely that the sagittal rays will be much affected by the vignetting, but the meridional rays certainly will be. The whole pattern of traced rays must be carefully fitted into the vignetted beam so that no unwanted rays are traced. When the design is complete, it is necessary to study the system to ascertain the clear apertures of all the mount details in order to achieve the specified degree of vignetting in the finished product. It is sometimes possible to specify the vignetting for two different obliquities, but there may be conflicting aperture requirements that will make this impossible.

E. PROGRAM LIMITATIONS

Optimization programs are generally written so that it is impossible to make a change in a lens that will increase the merit function, even though the next iteration will effect a large improvement. Also, no program will tell the designer that he should add another element or move the stop into a different air space. However, if an intelligent designer stops the program after a small number of iterations to see what is happening, he will quickly realize that an element should be divided into two, that the stop should be shifted, or that he should eliminate a lens element that is becoming so weak as to be insignificant. He may also decide to hold certain radii at values for which test glasses are available, letting the program work on only a few variables to effect the final solution. Also it is essential to remember that a computer optimization program will only improve the system that is given to it, so that if there are two or more solutions, as in a cemented doublet or a Lister-type microscope objective, the program will proceed to the closest solution and ignore the possibility of there being a much better solution elsewhere. It is this limitation that makes it very necessary for the operator to know how many possible solutions exist and which is the best starting point to work from.

Subject Index